Parks and Gardens of Britain

Parks and Gardens of Britain

A Landscape History from the Air

Christopher Taylor

Edinburgh University Press

For Stephanie

© Christopher Taylor, 1998

Edinburgh University Press
22 George Square, Edinburgh

Designed and typeset by Fionna Robson,
and printed and bound in Great Britain
by Redwood Books, Trowbridge, Wilts

A CIP record for this book is available from the British Library

ISBN 1 85331 248 7 (hardback)
ISBN 1 85331 207 X (paperback)

Contents

Preface

The origins of this book lie in a discussion many years ago with David Wilson, Curator of the Collection of the Cambridge University Committee for Aerial Photography; but it was not until 1996, when Richard Purslow, then of Keele University Press, recognised its possible interest, that the book was actually planned. David Wilson and his staff at Cambridge have provided me with every assistance while Rosemary Hoppitt, Elizabeth Whittle, Paul Everson, Keith Goodway, Paul Pattison, Bob Silvester, Robert Taylor, Angus Wainwright and Tom Williamson have all generously given me information on various gardens. I also wish to record my gratitude to all members, past and present, of the Historic Parks and Gardens Advisory Committee of English Heritage, as well as to the staff of the Parks and Garden Team of that organisation. All have unstintingly shared their expertise with me.

My greatest thanks, however, must go to my wife Stephanie. She introduced me to real gardens after I had spent half a lifetime looking at their archaeological remains, and she has encouraged and aided me in the writing of this book. The result is a personal and rather idiosyncratic view of parks and gardens by a landscape, rather than a garden, historian. The places chosen for illustration are, in the main, ones that I have either worked on or at least know well.

Christopher Taylor
Whittlesford, Cambridgeshire

Sources

All of the photographs used in this book are from the collection of the Cambridge University Committee for Aerial Photography, which holds the copyright. The identifying negative numbers, together with the dates on which the photographs were taken, are given at the end of each caption.

Bibliographical references in the text have been kept to a minimum, and those cited will often lead the reader to further studies. R. Desmond, *Bibliography of British Gardens* (Winchester: 1988), and G. and S. Jellicoe, *The Oxford Companion to Gardens* (Oxford: 1991), will also indicate further lines of enquiry. The standard works on gardens of particular periods and regions and general histories of gardens are given in the relevant places in the text.

For biographical details a number of standard works have been consulted without being specifically noted. These include: D. Banks and A. Esposito, *British Biographical Index*, four volumes (London: 1990); F. Boase, *Modern English Biography*, six volumes (London: 1965); *The Complete Peerage*, 13 volumes (London: 1910–53); *The Complete Baronetage*, five volumes (Exeter: 1900–6); *Dictionary of National Biography*, 80 volumes (London and Oxford: 1885–1996); *Who's Who* (London: 1849–1997); *Who Was Who 1897–1980*, nine volumes (London: 1981–96); *History of Parliament*, 25 volumes (London and Stroud: 1938–93).

The Publishers wish to acknowledge the generous award granted by the Marc Fitch Fund.

Chapter 1

Introduction

This book is a work of illustration, observation and interpretation of one facet of the landscape of Britain by a landscape historian. And, as it is by a landscape, rather than by a garden, historian, it is primarily about parks and gardens as they exist on the ground now. It has two themes. First, it is the history of landscapes designed for pleasure, seen through the existing and relict gardens that lie in the modern landscape. Second, it is about the people who paid for these landscapes and the motives that lay behind their creation. It is not about plants, except at a very simple level, for what is important about designed landscapes is their underlying structure. These structures may comprise 'soft' detail such as flower-beds, or 'hard' detail such as paths, walls or terraces. Alternatively, they can involve the ordered layout of the plants themselves as copses, avenues or bedding schemes. The basic elements can be further enhanced by the addition of decorative works such as pools, fountains or gazebos. It is this arrangement or structure that is the basis of designed landscapes. What distinguishes a seventeenth-century garden from a nineteenth-century one is not merely different plants but, more importantly, entirely different underlying structures and settings. And these settings are the result of differences in philosophy, politics, fashion, social attitudes, aesthetic appreciation and technology, as well as of plant collection and breeding.

As this book is about the differences in the structures of parks and gardens it is, in a sense, a book on archaeology. It seeks to interpret the physical evidence for the parks and gardens of Britain over 2,000 years and uses aerial photographs as illustration because such photographs give a different view of even the most familiar landscapes and thus enhance the understanding of their form and appearance. Aerial photographs, however, are not ideal source material for landscape studies and especially not for all gardens.[1] The primary difficulty is that they are not selective and show everything that is visible to the camera lens. Unless the viewer is experienced in using aerial photographs their interpretation is by no means straightforward. All features, shapes and periods of activity, whether significant or insignificant, can be seen and it is the observer who has to identify what is important. Another practical problem is that aerial photographs taken from an oblique angle, as are those used in this book, tend to 'flatten' the landscapes they are portraying. As a result, natural features that may have been important in controlling the original design, or in the way that the parks and gardens were viewed on the ground, can be difficult to appreciate from the air. Trees, although a vital part of most designed landscapes, also cause problems. Overall planting arrangements of trees are clear, of course, but by their very nature trees shade or obscure what is beneath them. Aerial photographs are also not very good at showing small gardens, partly as a result of restrictions on low flying, and this means that such gardens are underrepresented in any collection of aerial photographs. Even so, their value outweighs the disadvantages. The overview provided means that relationships not visible on the ground are clear and it thus gives insights and ideas about the development of gardens that might otherwise be missed.

There has, of course, always been an interest in such overviews. Even in medieval times small enclosed gardens were often located so that they could be seen from above, if only from adjacent buildings. Conway [5] and Nettleham [4] are examples. The same is true of the great medieval designed landscapes. In the late fourteenth century the mere at Kenilworth [10] was certainly intended to be viewed from John of Gaunt's first-floor hall. So too were the lakes and ponds at Leeds [11] from the late-thirteenth-century Gloriette. There was also a viewing platform or building overlooking the lakes at Leeds, as there was at Bodiam Castle, Sussex.[2] In the sixteenth and seventeenth centuries, the existence of terraces and mounts giving bird's-eye pictures of the gardens and the adjacent landscapes, as at Montacute [14], Tackley [19] and Wakerley [21], shows that the same tradition continued. The provision of roof-walks on contemporary houses,[3] and even 'monogrammed' water gardens which could only have been understood from the summits of adjacent mounts, as at Kettleby [16], exemplify this further. Many of the actual and idealised views of Elizabethan and Jacobean gardens are drawn as if viewed from above, as are the late-seventeenth-century depictions of Oxford and Cambridge colleges by Loggan and the eighteenth-century engravings of Kip and Knyff.[4] In the nineteenth century, the bedding arrangements below the walls and terraces of country houses were again intended to be viewed from above. Another value of aerial photographs is the way that they can show features not identifiable on the ground. In suitable light or weather, or in favourable crop conditions, soil or cropmarks can be seen, as well as slight undulations which, even if visible on the ground, are not readily comprehensible. The former flower-beds at Wakerley [21], Harrington [23] and Wimpole [49], as well as the layout of the former Cascade at Eastbury [78], are examples here.

However, perhaps the greatest value of aerial photographs, at least for the landscape historian, is the way in which they display the results of change in the landscape. The study of parks and gardens is the study of constant change, sometimes revolutionary and sometimes barely perceptible, but change none the less. And although these changes involve plants, designs of beds, forms of terraces or layouts of plantations, they also, perhaps more significantly, reflect changes in power, politics, aesthetics, economics, technology and social attitudes. In addition, the speed at which parks, and especially gardens, change makes them different from most other landscape features. A house, for example, will survive with annual maintenance. A garden left for a year will alter out of all recognition. Likewise, a building can be abandoned for decades and still be restored close to its original form. A garden neglected for the same time can rarely be returned to its earlier arrangement. The paths, terraces and ponds can be repaired or rebuilt, but the planting will have gone for good or have been altered irretrievably by natural growth. Further, for many old gardens the original plant types no longer exist. Even a relatively modern garden such as Sissinghurst [53] is no longer as it was when its original owners were alive. Time and circumstances have altered and continue to alter it.

The complete overview afforded by aerial photographs can often help to point to some of the changes, whether natural or man-made, that have taken place, but neither the photographs, nor any other form of historical evidence, can really answer the question most frequently asked about historic parks and gardens – what were they like originally? It is sometimes possible to get very close but never to see gardens 'as they were'. That is part of the frustration and the pleasure of garden history. Aerial photography for historical purposes has been undertaken regularly since the early 1930s. As a result, many of the photographs used in this book have themselves become historical documents. Many were taken in the 1940s and 1950s and show gardens in a state very different from how they are now. Some have been altered by restoration [62], some destroyed [79] and some modified by time and neglect [46]. That is, the aerial photographs themselves are now demonstrating change.

This book has been arranged in approximately chronological order to illustrate the development of parks and gardens from Roman times to the present day. The aerial photographs used to illustrate the discussion of parks and gardens dating from before the seventeenth century, and a few of later date, do not depict gardens at all but only the archaeological remains of former gardens. However, there is no alternative to using these archaeological gardens as evidence if the historian is to present a comprehensive story. Despite claims to the contrary, there are very few gardens earlier than the eighteenth century and probably none from before the seventeenth century that are, in any way, close to their original appearance. Most of those that are alleged or perceived to be seventeenth-century or earlier are either later recreations, as at Drummond [43], or are so fragmentary as a result of later alterations that they cannot be considered historically significant. Even those that seem to be convincingly original often lack documentation of sufficient detail for their original form to be certain. Levens [27] is an example of this.

The second theme of this book is that of human behaviour, for the history of parks and gardens has no meaning without an understanding of the people who created them. What makes designed landscapes, indeed all landscapes, fascinating to a landscape historian is that they are the visual expression of human individuality as well as being reflections of society as a whole. Landscapes without people are dead. Too many landscape histories do not include people and parks and gardens most of all demand an understanding of the people behind them. Of course, it is essential to identify the plants, planting, designs and structures, but without knowing the people who desired them, who made them what they were and from whence came their wealth, the understanding of parks and gardens is incomplete.

Few garden historians would disagree with this but the interests of many lie primarily with the plants, the designers and in identifying the broad cultural, aesthetic, philosophical or botanical themes that underlie them. The understanding of such themes is vital in garden history and the pages that follow reveal something about garden designers and political and philosophical trends. However, whether every garden or park was always the result of such trends may be doubtful, but whatever lies behind parks and gardens, in their various ways they usually convey something about their owners as well as about their designers. In the descriptions that follow it is the backgrounds of the owners that are stressed rather than those of the designers, although, in fact, these were sometimes the same. Inevitably this means looking at the upper levels of British society through the ages. This may be a somewhat unfashionable, or even politically incorrect, approach but it was the people from this level of society who created most of the parks and gardens that survive. To know these people, their motives and the sources of their wealth, allows a better understanding of them as individuals and of the society of which they were part.

It is not always easy to identify motives for human actions and in the case of parks and gardens it is often very difficult to establish the real reasons for their creation, but certain motives seem to stand out and are applicable to every age. First, there are the passionate gardeners who delight in plants and planting and whose primary interest is in the creation and enhancement of arrangements of plants. In this book the earliest such identifiable person is Queen Eleanor, wife of Edward I, whose interests are visible at Conway [5]. Other later people with the same passion for gardens include Colonel James Grahame at Levens [27], John and William Aislabie at Studley Royal [31], John Spencer at Cannon Hall [30] and the Nicholsons at Sissinghurst [53]. More often, perhaps, parks and gardens were laid out as status symbols, as statements of political allegiance or power, to enhance their owners' positions in society or to provide a backcloth for social activities. Examples here range from Quarrendon [18], created for an Elizabethan courtier, to Leighton [51], belonging to John Naylor, one of the richest non-aristocrats in Victorian Britain. Somewhat akin to these are the parks and gardens that are so crudely designed or simply laid out that one can conclude that their owners were only concerned with having roughly the same kind of surroundings to their homes as those of their equals or betters. Hackthorn [34] is an example of this. While the motives behind the gardens of Roman Britain, [2] and [3], will inevitably always remain unknown, they were really only pale imitations of classical Mediterranean ones created by a newly civilised people. There are also the plantsmen's gardens, usually filled with varieties of plants that their owners collected or bred. Some of these gardens have either a minimal underlying structure or none at all, as at Caerhays [60]. More often there is an ordered design that can sometimes be a sheer delight to the viewer, as at Batsford [52], and more clearly at Dyffryn [58]. In the majority of gardens, however, a combination of all these factors applies and one of the pleasures of landscape history is identifying them.

Finally, there are the landscapes designed for public, institutional or corporate access. The motives for these include the need to provide additional facilities to private housing estates, as at Calverley Park [63], to take into account advances in medical treatment as at Crichton [72], to meet educational and social concerns as at Christ's Hospital [70], Exeter University [71] and Cumbernauld [69], and to express civic pride, as at Cathays Park [67] and Princes Street Gardens [64].

The aerial photographs in this book thus show the evidence for the changing structures and arrangements which together make up the history of designed landscapes in Britain, as well as for some of the motives behind them. Collectively they illustrate the result of the continuous battles with nature that human beings wage. These are battles which neither has ever won or lost. The landscape emerges simply enhanced.

Chapter 2

The Beginning

Although strictly speaking the history of parks and gardens in Britain begins with the arrival of the Roman army in AD 43, it is of interest to glance briefly at prehistoric times. Agriculture was practised in this country from at least 3500 BC, but there is no evidence that prehistoric people had pleasure gardens. An argument often advanced to explain this lack of gardens is that they did not have time to spend on creating them. Certainly there were periods of warfare and famine when survival was the sole preoccupation, but the archaeological record is full of wonderful objects created at great expense by craftsmen of the highest order. These objects included jewellery, swords, shields and pottery which were often highly decorated. Prehistoric people also had complex ritual and religious activities and ceremonies on which even more time and resources were expended.

It is in connection with the religious needs of prehistoric people that some forms of 'gardens' may have existed. These were not necessarily man-made, but were perhaps only woodland glades, a riverine island, a spring or other water source, where communication, whether contemplative or barbarous, with gods might have taken place. This is speculative, but it seems likely that the first 'gardens' in Britain were prehistoric religious sites.

The evidence for this lies in the deposition, from around 1500 BC, of valuable objects in bogs, rivers or lakes. This was a practice that was widespread in Europe.[1] It seems to have involved a veneration of water and watery places and even, perhaps, an appreciation of a spectacular landscape as at Lyn Fawr **[1]**.

There is also literary evidence for the existence of sacred groves in late Iron Age times, such as those on Anglesey, destroyed by the Roman army in AD 60–1. Although the details of these groves are unknown, they may have involved some form of management of the landscape and thus have been a form of garden. One possibly related type of site at Thetford in Norfolk has been excavated. This site dated from the very end of the Iron Age and comprised a number of concentric enclosures partly filled with a dense pattern of posts. It has been interpreted as perhaps being an artificial version of the documented groves possibly used for meetings and worship.[2]

It was the incorporation of Britain into the Roman Empire that produced the first real gardens. These, however, were merely one aspect of the romanisation of a hitherto barbarian society which involved no racial changes. The Empire brought peace and access to both the material wealth and the intellectual traditions of the classical world. Within a generation, Iron Age warriors had become Roman country gentlemen and prehistoric peasants tenant farmers. Systems of local and national government appeared and a veneer of civilisation was applied. The most attractive, or useful, aspects of the Roman way of life, such as improved farming methods, industrial processes, urban life, writing, central heating and mosaic floors, were all quickly adopted. So too were gardens which were almost entirely Mediterranean in form and which, together with the new villas, were perhaps mainly intended to display the status of their owners. Most large town houses and almost all villas or country houses seem to have acquired gardens very quickly.

The details of these gardens are not entirely clear. Archaeologists in Britain have been slow to look for Roman gardens and although numerous town houses and dozens of villas have been excavated, few of these have assisted in the understanding of the surrounding gardens. Those discoveries which have been made have usually been interpreted in purely functional terms. Paved areas, timber structures, ponds and walls have all been found adjacent to often palatial villas, but have then been identified as yards, pens, drinking ponds or agricultural buildings, and not, as perhaps would be more likely, as paths, terraces, pools and gazebos. Only recently have archaeologists begun to look for, and to find, gardens.

The best find is at the palace-like villa at Fishbourne, Sussex, where excavations have revealed a large formal garden. This, because it occupied a courtyard linking the main entrance to a sumptuous audience chamber, was clearly meant to impress visitors. It was 75 metres by 100 metres in extent, surrounded by a veranda and with axial and encircling paths, probably edged with box. Traces of timber gazebos and the water-supply system for fountains or basins were discovered. There was also a separate, more informal, garden overlooking the seashore with a pool and a stream.[3] At other villas, similar but less complete traces of gardens have been found. Paths and flower-beds are known at a villa at Frocester Court, Gloucestershire, while a large formal pond and a probable gazebo were discovered at a villa at Milton Keynes.[4] Gardens have also been found in Roman towns. In London, a large, richly appointed building, probably a *praetorium* or the provincial governor's residence, had a pool and two fountains in one of its courtyards.[5] All this shows that gardens in Roman Britain were very close in form to those known from the Mediterranean world, from both excavations there and from the classical writers. It is thus possible to appreciate what must have existed at villas such as Lidgate [2] and grand urban buildings, such as that at Silchester [3], which are now only dimly visible as cropmarks on aerial photographs.

The existence of informal gardens with views, as at Fishbourne, is also known from classical writings. Pliny's Laurentian Villa had an enclosed garden but mountain and countryside views as well, while others had vistas through trees, perhaps similar to late-seventeenth- or early-eighteenth-century parkland.[6] There is also archaeological evidence that many large continental villas in later Roman times, especially in France, had extensive designed landscapes as well as formal gardens.[7] As yet, totally convincing British examples remain to be found, although it has been suggested that a villa at Rivenhall, Essex, lay at the centre of an intricate formal landscape of woodland, avenues and vistas.[8] This brief flowering of exotic gardens in Roman Britain ended in the fifth century as the protective Roman army withdrew and British society reverted to a less civilised way of life. The Dark Ages had begun.

1 Lyn Fawr, Glamorgan

This is not a garden, nor even a man-made landscape, but it may hint at the beginning of an appreciation of landscape in Britain. Lyn Fawr stands on the northern edge of the South Wales valleys where the gently sloping sandstones terminate in a dramatic scarp over 300 metres high, overlooking the valley of the River Neath. The great amphitheatre-shaped corrie was formed by snow and ice during the last Ice Age. In 1911 a small, natural lake at the base of the steep slope was drained prior to the construction of the reservoir that now occupies the site. During the work various objects were discovered dating from approximately 650 BC. They included axes, sickles, gouges, scabbards, a razor, two cauldrons and some horse-harness, all of bronze, and a sword, a sickle and a spear of iron. Pieces of shaped wood were also found.

A detailed analysis indicated that while some of these objects were locally made, others, such as the razor and probably the axes, came from south-west England, while the sword is unique in Britain. Although little in prehistory is ever certain, the most likely explanation for these finds is that they were part of a votive offering thrown into the lake as gifts for, or as propitiation to, the gods who dwelt there. The timber would have been a platform on the side of the lake on which rituals were conducted and from which the objects were thrown.

The origin of some of the objects has led to the conclusion that they were part of the spoils seized by local hill-men during raids on the lowlands. After a successful attack on the wealthy inhabitants of the lush valleys of the River Severn, Somerset or Devon, these late Bronze Age warriors would have returned home in triumph. As thanksgiving for their achievements they would have given part of the treasures they had brought home to their gods who lived on or below the slopes of the mountain. The belief that Lyn Fawr was the home of their gods may well have stemmed from its throne-like nature, but also, perhaps, from an appreciation of landscape form.

H. N. Savory, 'Some Welsh LBA hoards', *Archaeologia Atlantica* 1.2 (1976), p. 122; H. N. Savory, *Guide Catalogue of the Early Iron Age Collections*, National Museum of Wales (Cardiff: 1976), p. 21; neg. no. CGB 14, 19 June 1978.

2 *Lidgate Villa, Suffolk*

Here is no certain garden, perhaps rather the setting for one, visible only as soil and cropmarks from the air. What can be seen, between the lines made by modern farming methods, are the remains of a large Roman villa or country house, at Lidgate, south-east of Newmarket. The site has never been excavated and although it seems to be all of one period, examination would no doubt reveal that, as is the case with so many similar villas in Britain, it underwent many changes and adaptations throughout its life. What is visible probably represents the arrangement of the house and its surroundings at its greatest extent, perhaps in the fourth century AD. It lies on clayland, on a south-south-west-facing slope in a field where ploughing produces quantities of tiles and other building debris. The view here is from the south.

The main house is U-shaped in plan with a principal range some 50 metres long and with two short projecting wings. The individual rooms are visible, as is the corridor or veranda which extends around the building on the south. It faces on to what was presumably either a courtyard or, more likely, a formal garden. The necessary domestic or agricultural outbuildings are some distance away, making it likely that this courtyard was indeed a garden although there are no indications of any interior layout. The large building to the south-west, some 25 metres long and with buttressed walls, is probably a barn and there are other structures to the south.

J. K. S. St Joseph, 'Air reconnaissance: recent results 35', *Antiquity* 45 (1971), pp. 224–5; neg. no. BFD 47, 24 April 1971.

3 Silchester, Berkshire

Silchester is one of the best understood Roman towns in Britain, largely because, unlike most others, it was not reoccupied in later times. As a result, it was extensively excavated between 1890 and 1909. Although these excavations were successful by the standards of their time, much of the detail that archaeologists would now seek was ignored or destroyed. This included the evidence for urban gardens, although the remains of dog rose, mallow, violets and box were found.

Aerial photographs of Silchester are also valuable in that all the streets and the foundations of most of the stone buildings are visible as cropmarks, even though the more numerous timber buildings cannot be seen. This photograph shows the outlines of a splendid building which was either a mansio or inn where official visitors stayed, or possibly a praetorium which the provincial governor of Britain might have occupied on his progresses.

It is dated to the late first or early second century AD and lies on the south side of the town close to its walls. The building comprises a main block with two wings set around a courtyard. This main block appears to have held the principal reception rooms, including an apsidal-ended space which could have been the governor's audience chamber. The main entrance was through the right-hand wing where the metalling of a side road leading to an elaborate portico is clearly visible. The large central courtyard (45 metres by 35 metres) was thus not an entrance but an enclosed area to be viewed from the adjacent buildings. Although there is no proof, the probable high status of this building makes it likely that there was an elaborate formal garden here. It perhaps had a central fountain or sculpture and was divided into box-edged borders, an arrangement known from elsewhere in Roman Britain.

G. C. Boon, *Silchester, the Roman Town of Calleva* (Newton Abbot: 1974), p. 138; J. Wacher, *The Towns of Roman Britain* (London: 1975), pp. 262–4; neg. no. YN 56, 19 June 1959.

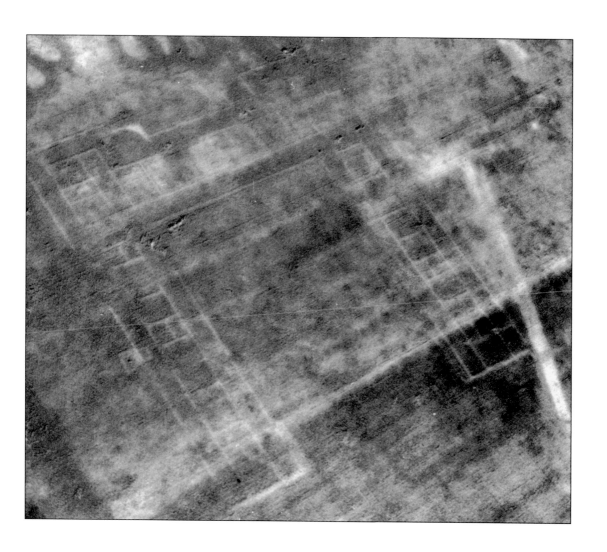

Chapter 3

Medieval Gardens

The period between the withdrawal of the Roman army in the early fifth century and the arrival of the Normans in 1066 is often called the Dark Ages. For most historians these centuries are not as dark as they used to be, for modern scholarship has shed new light on them; but for garden historians they remain murky. Almost nothing is known of gardens in Britain at that time beyond the fact that they certainly existed. Throughout the period Britain remained in contact with its continental neighbours through trade and religion. There was much interchange of people, goods and ideas with the rest of western Europe, the Mediterranean and even beyond, and in these places there is considerable evidence for gardens.[1] Certainly, knowledge of gardens and garden layouts was widespread in Britain by the ninth century and by the tenth century gardens are recorded, if rarely described in detail.

From the twelfth century, there is a steadily increasing amount of written evidence for pleasure gardens in Britain. These include descriptions, and details of payments for plants, construction and maintenance work. By the fourteenth century the evidence is considerable and is reinforced by similar material from the continent of Europe as well as by illuminated manuscripts and paintings from there.[2] Until recently these sources led garden historians to believe that most medieval gardens in Britain were relatively small, totally enclosed by walls or hedges and contained formal layouts with raised beds, arbours, fountains and paths. This is certainly what is usually depicted in the many continental illustrations and in the few British examples.[3] The monastic tradition of using enclosed cloisters for contemplation and prayer also appeared to give additional support to this belief even though cloisters were not gardens in a real sense.

Over the last few years, however, landscape historians and archaeologists have discovered the physical remains of medieval gardens and these have changed the accepted view of what they were like. Certainly some of these newly discovered gardens are just as medieval ones were thought to be. The fourteenth-century garden of the bishops of Lincoln at Nettleham [4] is only 45 metres by 60 metres and, originally surrounded by a high wall, it has the remains of paths, flower-beds and low terraces and was once overlooked by the bishop's palace. Another, created for Queen Eleanor at Conway Castle [5] in the late thirteenth century, is similar although it is squeezed behind the castle walls. A garden at Tintagel, Cornwall, is also a small walled enclosure[4] as are the monastic gardens at Haughmond Abbey [6], but there are others which are very different. The garden of the archbishops of York at Cawood [7], certainly in existence by the mid-thirteenth century, is bounded by a water-filled ditch or moat and contains three formal long ponds and two smaller ones as well as paths. Even double-moated medieval gardens are known. The early-fifteenth-century Pleasance at Kenilworth [10] is one such example. Another, probably of fourteenth-century date, is at Linton, Cambridgeshire while a third exists at Beckley Park, Oxfordshire.[5] One similar to these and dating from the early fourteenth century survived at Peterborough until the eighteenth century.[6]

Different again is the garden belonging to the bishops of Ely at Somersham [8], part of the boundary of which was a small stream set in an artificial, informal, 'vale'. And, unlike the older perception of medieval gardens, Somersham was large, covering some 14 hectares, considerably in excess of, for example, the alleged extensive royal garden in Gloucester which covered only six hectares.[7] Another large medieval garden was that at Spaldwick, Cambridgeshire, belonging to the bishops of Lincoln, which seems to have been bounded by a large earthen bank, perhaps surmounted by a hedge.[8] The interiors of these gardens also exhibited great variety. Somersham had raised terraced walks, a moated gazebo, pools and a line of ponds which increase in size thus counteracting the effect of perspective. The walled garden of the bishops of Winchester at Bishops Waltham, Hampshire, seems to have contained a meandering stream as well as a pond by the late fifteenth century.[9] With hindsight it can now be seen that both documents and illustrations actually record all these features now recognised by archaeologists.

The same is true of some very different medieval gardens recognised in Britain only recently. These have been called 'designed landscapes', an accurate if not ideal description.

What have been discovered are large areas of land which have been modified by the construction of lakes, ponds, embankments and drives, and almost certainly by planting, to form something similar to the landscaped parks of the eighteenth century. As yet, the origin and exact development of these medieval landscapes is not clear, for although over forty of them are now known, they vary considerably in detail. Most seem either to have been based on an earlier deer park or to have had a deer park constructed as part of their arrangement. Medieval deer parks have their own history, quite separate from that of medieval designed landscapes. They certainly existed in late Saxon times but they became popular under the Norman kings and by the thirteenth century there were hundreds of them in Britain,[10] varying greatly in size from a few hectares in extent to many kilometres across. Most were enclosed by earthen banks surmounted by wooden pales or, more rarely, by walls. These parks remained popular and useful to many members of the upper ranks of medieval society until the fourteenth century. One of them, Devizes [9], is illustrated not merely as an example but to indicate that, with its central pond and keeper's lodge and the pivotal and elevated position of the castle, it has some of the characteristics of designed landscapes. The parks of these landscapes also usually surrounded, or extended from, a castle or a principle residence of their owners or contained an often luxuriously appointed viewing tower, hunting lodge, or even temporary bowers.[11] However, what sets medieval designed landscapes apart from mere deer parks are features that made them pleasurable to look at and able to accommodate entertainments other than hunting. These features varied from simple additions, such as the thirteenth-century lakes overlooked by lodges for accommodating both huntsmen and their ladies, as at the deer parks at Harringworth, Northamptonshire, and at Kelsale, Suffolk, to complex and extensive formal gardens around a palace for which the deer park was merely a backdrop, as at Somersham [8].[12]

All levels of medieval aristocratic society seem to have been involved in the making of these landscapes. The Crown created many, including those at Kenilworth [10], Leeds [11] and Bolingbroke [12], as well as at Odiham, Hampshire.[13] Most included lakes of considerable size, the principal functions of which may have been to provide for activities such as boating, fishing and perhaps water pageants, as well as pleasing views. The greater aristocrats and leading clerics did much the same. The earls of Norfolk at Framlingham [13] and the earls of Arundel at Clun, Shropshire, created lakes or meres within parks at their castles.[14] The bishops of Lincoln laid out complex landscapes at Stow, Lincolnshire, as early as the twelfth century, and at Lyddington, Rutland, as did the bishops of Ely at Somersham [8].[15] The bishops of Worcester had one at Alvechurch, Worcestershire, and those of Winchester at Waltham, Hampshire. Even rich merchants such as Laurence de Ludlow at Stokesay, Shropshire, could have a designed landscape. Perhaps the most remarkable, although curiously without a park, is at Bodiam Castle, Sussex, where, in the 1380s, Sir Edward Dalingridge surrounded his new castle there with a landscape of moats, lakes, ponds, cascades and terraces as well as with a hill-top lodge from which to view it all.[16]

Now that these landscapes have been recognised on the ground it is obvious that they are also recorded in documents. Henry II certainly had one at Woodstock, Oxfordshire. Not only did the park there have a menagerie containing exotic animals, it also had great lakes and an isolated lodge where the King kept his mistress, Rosamond Clifford. The lodge had at least two gardens with pools.[17] Many continental paintings and illuminated manuscripts also seem to show contrived landscapes, either alone or as backdrops to formal gardens. Particularly important is that some appear to show arrangements of trees deliberately planted in lines, avenues or copses, and apparently managed. While these illustrations are usually imaginative or idealistic they probably reflect reality. Indeed, designed landscapes appear to be recorded on the continent as early as the ninth or tenth centuries. Perhaps the greatest was the late-thirteenth-century one of the dukes of Burgundy at Hesdin. This too had gardens and a park below the castle but the park also contained separate gardens, lakes and a lodge.[18]

Although no actual illustrations of designed landscapes exist for Britain, a few early maps have tantalising suggestions. A 1607 plan of Windsor Castle shows it and its formal garden lying at one end of the Little Park. More significantly, the park is depicted with an avenue and with two artificial alignments of trees which may have been created using former hedgerows as in some eighteenth-century parks.[19] The Windsor arrangements may have been very late medieval in date, for designed landscapes continued to be created, as can be seen by the lakes at Baconsthorpe Castle, Norfolk, and the ponds and parkland at Collyweston, Northamptonshire, both of late-fifteenth-century date.[20]

As with all parks and gardens, the reasons for these medieval ones are varied and complex. Those at Conway [5] and at Leeds [11] were perhaps the result of the love of gardens by their royal owner. It is difficult to be certain that such interests lay behind other medieval parks and gardens although there are hints. Dalingridge, who was a retired soldier, civil servant and politician, was perhaps indulging his fantasies when he created the Bodiam landscape. And although the motives of the bishop of Lincoln responsible for Stow, Lincolnshire, are unknown, the designed landscape there was appreciated for its aesthetic qualities as early as the 1180s. Giraldus Cambrensis, who knew it then, described the palace there as 'delightfully surrounded with woods and ponds'.[21] Further, the importance of contemplation and the desire for tranquility in what was a brutal world comes through in contemporary writings and illustrations, both secular and religious. More obvious, however, is the evidence which points to a desire for ostentatious display and expressions of status and political and economic power. The elaborate landscape at Clun Castle, Shropshire, seems to have followed immediately the elevation of the Fitz Alans to the earldom of Arundel while the landscape at nearby Stokesay can hardly be anything less than an attempt by the richest merchant in England to imitate what he had seen while working for his king.[22] At Nettleham [4], too, the garden was certainly used for more than social activities. In 1432 it was the scene of a formal consultation between Bishop Grey and the Dean and other dignitaries of Lincoln Cathedral.

The desire to impress visitors to medieval gardens and designed landscapes is also seen in the way they were intended to be approached. At Woodstock, Stow and at Somersham [8] the approach drive to the palaces passed along causeways set between large lakes which certainly at Somersham would have given the impression of buildings rising out of water. This was only achieved at Somersham by the removal of the adjacent village and its relocation to a new site.

4 Nettleham, Lincolnshire

This is the site of one of the many palaces of the medieval bishops of Lincoln. Nettleham is three kilometres north-east of Lincoln and thus was used to accommodate dignitaries visiting the cathedral. It was also a favourite residence for many of the bishops. The palace was abandoned in the sixteenth century and was demolished soon afterwards. The view here is from the south-east with the village of Nettleham in the background. Only part of the remains of the actual palace is visible. Most of it lay in the area of the existing houses below the street, top right. However, the uneven ground immediately to the left of the large building close to the street marks the position of the private apartments of the palace.

The embanked approach drive can be seen in the foreground. It enters an outer courtyard, passing between the foundations of two rectangular buildings, probably barns. The large hole on the left-hand side of the courtyard is a chalk quarry, the dimpled mound beside it being a medieval limekiln. Between the courtyard and the street and to the left of the palace is a garden. It is roughly rectangular in shape and, to judge from the stony bank around it, clearest in the top left-hand corner, it was bounded by a high wall on three sides and visible from the palace on the fourth. Within it are traces of low terraces, some raised footpaths and flower-beds. Excavations on the site in 1959 revealed rich, dark soil adjacent to one of the paths, presumably from a flower-bed. This garden is thus exactly the type that appears in many late-medieval illustrations and in contemporary documents.

Its medieval date is not in doubt. It was laid out in about 1336 when Bishop Burghersh was licensed to crenellate his manor and surround it with a wall. That the garden was created at this time was confirmed by the archaeological excavation which produced a well which had been filled in and then covered by the boundary wall, an event dated to the mid-fourteenth century.

P. L. Everson, C. C. Taylor and C. J. Dunn, *Change and Continuity* (London: 1991), pp. 129–31; neg. no. EX 3, 3 July 1950.

5 Conway Castle, Caernarfonshire

Here is an apparently unlikely place for a garden. Conway Castle, together with the adjacent fortified town, was built by Edward I between 1283 and 1287 as a stronghold against the Welsh. It guarded the crossing-place of the River Conway and demonstrated the power of the English crown over the ancient kingdom of Gwynedd. The great towers of the castle, for linear protection rather than point defence, were the most up-to-date in Europe at that time. The view here is from the east and at the extreme eastern tip of the castle, just above the river, is a small walled court protected by three low towers. Militarily this was the East Barbican which gave access to the Water Gate, but it was also a garden which could be viewed from the main South-East Tower above it where the royal sleeping apartments were situated, or from the Privy or Presence Chamber set between the two principal towers.

The inspiration behind this garden was Edward's first wife, Eleanor of Castile. She spent much time with her husband and was a committed gardener who may have introduced the hollyhock to England from Spain. She had a number of gardens or landscapes created for her including ones at Leeds Castle [11], at King's Langley, Hertfordshire in 1279–81, and others at Rhuddlan and Caernarvon Castles, also in the 1280s. The garden at Rhuddlan is well documented. It too was in a walled courtyard laid with turf and with a central pond surrounded by seats. The one at Conway may well have been similar and certainly contained a lawn edged with staves from discarded casks. It survived until the seventeenth century and a painting from 1600 shows it divided into four knots bounded by paths.

R. Allen Brown, *Castles from the Air* (Cambridge: 1989), pp. 97–100; H. M. Colvin, *History of the King's Works* I (London: 1963), pp. 341–52; E. Whittle, *The Historic Gardens of Wales* (London: 1992), pp. 9–10; neg. no. BUB 86, 7 July 1975.

6 *Haughmond Abbey, Shropshire*

Haughmond Abbey lies five kilometres north-east of Shrewsbury. Although ruinous, it has both enclosed cloisters, which were not strictly gardens, and the private gardens of the abbot and the prior. The early history of Haughmond is unclear. By 1135 it was in existence as a house of Augustinian canons and was richly endowed, largely through the patronage of the Fitz Alans, lords of Oswestry and Clun and later earls of Arundel. As a result, the abbey enjoyed considerable prosperity. It was suppressed in 1539.

The view here is from the north-west. The arrangement of buildings is unusual, partly as a result of the steeply rising ground to the east. In the left foreground the outlines of the late-twelfth-century monastic church are visible. Above it is the main cloister with a large tree on its north side and the chapter house on its east. On the far side of this cloister are the foundations of the refectory and beyond that, unusually, is a 'little cloister'. This has the remains of the kitchen range on its west, right, and the dormitory on the east. On the south is the upstanding fourteenth-century abbot's hall with its private apartments, including a great late-fifteenth-century south-facing bay window, to its left. This window overlooks a court, the western part of which was almost certainly the abbot's private garden.

Another garden, known as Longnor's Garden after Abbot Nicholas de Longnor (abbot 1325–46) who presumably created it, occupied the walled enclosure to the east of the dormitory range. In the mid-fifteenth century this range was converted into the private apartments of the prior and a door from them led into the garden. As with the abbot's garden its layout is unknown but it certainly contained a dovecote. Although the church and the dormitory range were demolished soon after the dissolution, the abbot's lodgings, as often happened, were turned into a private house and occupied until the mid-seventeenth century. During this period the 'little cloister' was used as a formal garden.

N. Pevsner, *Shropshire* (London: 1958), pp. 140–3; G. Chitty, *Haughmond Abbey: a Brief Guide* (London: 1992); VCH, *Shropshire* II (London: 1973), pp. 63–70; neg. no. ES 76, 14 May 1950.

7 Cawood Palace, Yorkshire, West Riding

The site of another medieval ecclesiastical garden, viewed here from the south. It was attached to a palace of the archbishops of York which was used regularly from the thirteenth to the sixteenth century, after which it was abandoned. The location near the junction of the Rivers Ouse and Wharfe allowed easy access to York.

The palace itself stood south of the river in the area of modern housing below the bridge. It was arranged around two courtyards, although all that now survives is a magnificent gatehouse and adjoining range, built by Archbishop Kemp in 1444–5. This is visible above and right of the clump of trees in the centre middle distance. The open land in the foreground which lay to one side of the palace entrance had a variety of uses. The long curved hollow in the left foreground was a medieval dock, linked to the river by a canal. The indistinct depression adjacent to and north of this dock is an old clay-pit, probably the site of the archbishop's tile-works, recorded in 1235 as abutting on the ditch of his garden. This ditch or moat is in the centre right of the photograph bounding a rectangular enclosure which is the garden itself. At the south, nearest, end of this garden are the remains of three long parallel ponds, one edged by trees and still full of water, the others, now dry, to its right. Until recently there were two other ponds at the far end of the garden, since destroyed. In the centre of the garden are two blocks of slight narrow ridges, separated by a pathway running axially down the garden. The ridges may be for the planting of an orchard as the area was perhaps called Apulgarth in the early sixteenth century. A recent archaeological resitivity survey has detected various circular and linear features in the garden which must also relate to the medieval layout.

N. K. Blood and C. C. Taylor, 'Cawood: an archiepiscopal landscape', *Yorkshire Archaeological Journal* 64 (1992), pp. 83–102; M. A. Cole et al., 'Non-destructive techniques in English gardens: geophysical prospecting', *Journal of Garden History* 17.1 (1997), pp. 29, 31; neg. no. ZX 78, 8 July 1959.

8 Somersham Palace, Cambridgeshire

This is another garden of a medieval bishop but very different from both Nettleham **[4]** and Cawood **[7]**. The photograph, taken from the south, shows part of the ploughed-out remains of this garden as cropmarks. Its significance, however, is that it was only a small part of an extensive designed landscape probably laid out in the thirteenth or fourteenth century. Somersham was acquired by the bishop of Ely in 1109 and from the late twelfth century it became an important episcopal residence. It was regularly used by the bishops in the thirteenth and fourteenth centuries and was visited by a number of medieval kings. By the fifteenth century it was rarely used and by the sixteenth was ruinous.

The palace was approached from the north past the parish church and along a drive created by the removal of the medieval village of Somersham which was relocated over its own fields further north. The drive crossed an embankment between two large lakes and entered an extensive moated area, with a garden, within which the double-courtyarded palace stood. The bed of one of the lakes, the rectangular open area in the centre background below the church, is still visible, as is the roughly triangular moated site below again. The farm buildings in the centre stand on the site of formal gardens outside the moat, with the trees to the right hiding a line of ponds.

In the foreground are the traces of a much larger garden, enclosed by banks and watercourses. The straight white marks are the foundations of former long, raised terraces, the dark marks are former ponds. In the left-hand corner is a tiny ploughed-out moated site with adjoining ponds, perhaps a pleasance. To judge by the pottery and tiles from its interior it enclosed a fourteenth-century summer-house. A similar garden, with more ponds, lay to the right, off the photograph, beyond the straight farm track. The latter, a raised causeway overlooking the gardens on each side, originally gave access to a large deer park which occupied 250 hectares of rising ground to the south. From there the whole palace, moat, lakes and gardens would have been visible as a bird's-eye view.

C. C. Taylor, 'Somersham Palace, Cambridgeshire: a medieval landscape for pleasure?' in M. Bowden, D. Mackay and P. Topping (eds), *From Cornwall to Caithness*, British Archaeological Reports (British Series) 209 (1989), pp. 211–24; neg. no. BTI 22, 26 May 1975.

9 Devizes Park, Wiltshire

Here is a typical medieval deer park. It was laid out, probably in the early twelfth century, by Bishop Roger of Salisbury (bishop 1107–39), who also refurbished his castle at the same time and perhaps laid out the new town of Devizes to the north of it. Roger was Henry I's great Justiciar and the organiser of his Exchequer. The castle was seized by Stephen in 1139 and Henry II later decided to retain it formally in the hands of the Crown. It and the park remained Crown property throughout the medieval period and both were used as part of the dower of successive queens. From the beginning of the fourteenth century the castle became a residence rather than a fortress and the park formed an important part of its amenities.

The castle lay at the extreme north-eastern end of the park and from its walls most of the two square kilometres of the park could be seen. In 1157 the park was bounded by a massive bank and internal ditch, both of which still survive over most of its circuit. In the early thirteenth century this bank was surmounted by a stone wall. Thereafter a wooden fence or pale took its place. The deer within the park were managed by a keeper who lived in a moated lodge in its centre alongside a fishpond. The park was disused by 1595 and was subsequently divided into the fields that still exist.

The view here is from the south-west. In the far distance is the town of Devizes, with the castle hidden in the clump of trees below it, right of centre. The continuous curving boundary of the park can be seen running south-west from the houses near the top left-hand corner. It meets and follows the road in the foreground until the latter turns sharply. The park boundary continues, curving south-east and then east, right, until it meets a sinuous road running north-east which it follows until it reaches the castle. The small clump of trees in the middle distance, just left of centre, is the site of the moated lodge, the ditches and fishpond of which were fed by the stream visible in the foreground.

M. W. Beresford, *New Towns of the Middle Ages* (London: 1967), p. 504; H. M. Colvin (ed.), *History of the King's Works* II (London: 1963), pp. 626–8; VCH, *Wiltshire* X (Oxford: 1975), pp. 245–6; neg. no. UO 14, 15 April 1957.

10 The Pleasance, Kenilworth, Warwickshire

The Pleasance is a royal garden retreat, but also only a small part of an extensive designed landscape. This landscape was perhaps begun in the early thirteenth century when a small artificial lake in the valley to the south of Kenilworth Castle was enlarged to become a vast mere, covering some 450 hectares. This work was possibly by King John (1199–1216) in the early thirteenth century. The mere was constructed mainly for military purposes but also undoubtedly to enhance the setting of the castle. This setting came into its own in 1279 when the castle was held by Edmund, Earl of Lancaster (1245–96), younger son of Henry III. Under the presidency of Roger Mortimer (?–1285) one hundred knights and their ladies assembled for tilting and tournaments. This use of the castle and its mere was continued in 1361 when Kenilworth passed to John of Gaunt, Duke of Lancaster (1340–99). The improvements then made to the castle included a remarkable first-floor hall, from the window seats of which the whole of the mere and no doubt the boats thereon could be viewed.

In 1399 the accession of John of Gaunt's son, Henry Bolingbroke, as Henry IV, returned Kenilworth to the Crown. It was his son, Henry V (1413–22) who, in 1417, created the 'Plesauns en Marys' here seen from the north-west. As its name suggests, it was reclaimed from the marshy edge of the mere and comprises two concentric rectangular moats, separated by a flat-topped terrace walk. The outer moat was linked to the mere by a short length of canal visible in the left background, indicating that its main approach from the castle was by water. As can be seen from the slight foundations still visible, the walled interior contained one large stone building on the right-hand side set within a rectangular enclosure, perhaps the documented hall and associated chambers. There were also corner towers. Then, or perhaps later, there was also a timber banqueting house which was removed in the early sixteenth century when the site was abandoned. Other slighter marks may be the traces of flower-beds. The mere survived to become, in 1573, part of a background to another splendid occasion, the last visit of Elizabeth I to Kenilworth, then held by her favourite, Robert Earl of Leicester. It was finally drained in the seventeenth century.

H. M. Colvin (ed.), *History of the King's Works* II (London: 1963), pp. 682–5; M. W. Thompson, *Kenilworth Castle*, English Heritage Guide (London: 1991); M. W. Thompson, 'Reclamation of waste ground for the pleasance at Kenilworth Castle', *Medieval Archaeology* 8 (1964), pp. 222–3; neg. no. AMX 91, 30 January 1966.

11 Leeds Castle, Kent

Lord Conway (1856–1933) called Leeds 'the loveliest castle in the world'. He was referring, of course, to its present appearance, but there is evidence to suggest that it was designed in the late thirteenth century to be just that.

The old castle at Leeds, about which little is known, came into the hands of Edward I in 1279 and both castle and manor were held by Edward's first wife Eleanor of Castile until her death in 1290. Leeds was among the most favoured of residences for Edward and his queen, evidenced by the vast amounts of money spent on it in the 1280s, as well as by the numerous royal visits there. By the 1270s the castle, seen here from the south-east, probably consisted of a motte and bailey corresponding to the main island and detached tower and was set within a great deer park. The main island was then rebuilt and given new defences and the great surrounding lake, together with another to the south, was probably part of this work. That this is so is suggested by the fact that the detached tower was built, much as it is now, by Edward between 1279 and 1288, but not as a defensive structure. It was designed from the beginning to be, and was called, a 'Gloriette', a term originally meaning a summer-house but later coming to signify a lodge of lordly apartments within a park or garden. That is, it was meant to be a place from which the park, lake and other features could be seen and admired. Among these other features were a set of small curving ponds, now dry, extending up the hillside to the east, right, of the Gloriette, as well as larger ponds to the north-west. There is also on the hillside west of the castle the site of another, perhaps contemporary, viewing platform from which the castle, lakes and park could be seen as if from the air.

Leeds remained mainly in royal hands until the sixteenth century and its landscape underwent a number of improvements including an enlargement of the park, the incorporation of another park and, in 1369–73, the building of a new lodge. The present parkland is mainly of the eighteenth century in appearance and may have been laid out by Capability Brown.

D. A. H. Cleggett, *History of Leeds Castle and its Families* (Leeds: 1992); H. M. Colvin (ed.), *History of the King's Works* II (London: 1983), pp. 695–703; neg. no. BBP 62, 26 May 1970.

12 Bolingbroke Castle, Lincolnshire

Another, but different, medieval designed landscape is at Old Bolingbroke ten kilometres south-east of Horncastle. The original Norman castle lay on a hilltop 500 metres north, left, of this site. It was apparently abandoned and a new castle constructed in 1220–30, on a completely new valley-bottom position, by Randulph de Blundevill, Earl of Chester and Lincoln (1181–1232). Whether this new castle was deliberately sited on level ground in order for it to have a designed landscape is not known but its position is tactically weak. The setting could also have been created in the late thirteenth century when the castle was the administrative centre of the estates of Henry de Lacy, Earl of Lincoln (?–1311). Or it may have been made in the later fourteenth century when Bolingbroke came, first by marriage into the hands of the earl of Lancaster, and then through the last of the Lancaster heiresses to John of Gaunt (1340–99), son of Edward III. Certainly his son, Henry Bolingbroke (1367–1413), later Henry IV, was born here and John of Gaunt seems to have been appreciative of the landscape at Kenilworth [10]. The castle had much money spent on it in the early fifteenth century but by the 1460s maintenance had dwindled and its great days were over.

The view here is from the west. The polygonal foundations of the castle and its dry moat are clear. To its south, right, a large rectangular embanked area, with modern drainage ditches crossing it, was originally an extensive shallow lake, the water-level of which was actually higher than that of the ground outside. Traces of further small ponds are visible to the east and west of the castle and along the eastern edge of the lake. Except from the north, which was the main approach, all views of the castle would have been reflected in sheets of still water. The large embanked moated pond in the centre of the former lake is a more recent feature. The lake was certainly drained by 1718 when its bed was used as an animal pound. The pond was added later, perhaps for watering stock.

M. W. Beresford and J. K. S. St Joseph, *Medieval England: An Aerial Survey* (2nd edn, Cambridge: 1979), pp. 50–2; H. M. Colvin (ed.), *History of the King's Works* II (London: 1963), pp. 571–2; M. W. Thompson, 'The origins of Bolingbroke Castle', *Medieval Archaeology* 10 (1966), pp. 152–8; neg. no. AMU 94, 3 January 1966.

13 Framlingham Castle, Suffolk

The evidence for a medieval garden at Framlingham Castle is inconclusive but there was certainly a designed landscape there. The first castle, a simple motte and bailey, was probably built by Roger Bigod (?–1107). It was rebuilt in the mid-twelfth century by Hugh Bigod (c. 1095–1176), perhaps on his elevation to first earl of Norfolk in 1141, and rebuilt again, virtually as it is in the photograph, in 1190 by Roger (?–1221), the second earl. It remained in the hands of the Bigods until 1306.

It was probably at some time in the late twelfth or early thirteenth century that one of the earls created the designed landscape here which comprised two parts. The first was a deer park of 300 hectares, first recorded in 1270, which, with the castle at its southern end, stretched north, left, for some three kilometres. Within the park was a lodge which itself had a garden by the sixteenth century. The second feature comprised two large artificial lakes or meres immediately below and to the north of the castle, formed by the damming of a tiny stream and also there by 1270. The southern mere, once covering some 9.4 hectares, still exists, although now only a shadow of its former self, and is visible as marshy ground in the photograph which is taken from the west. Only the dam of the second, northern, lake survives. A garden at Framlingham is first recorded in 1302 and probably occupied the rectangular court which lies between the castle walls and the mere. This garden is otherwise undated and only directly documented as such in the sixteenth century when Thomas Howard (1537–72), the second duke of Norfolk, retired to Framlingham. With its flat interior and outer terraced walkways it is typical of that date and in the early nineteenth century still contained two small ponds, but as a structure it must be much earlier, for the walls that project north-west from the castle to meet the terrace walks are of the 1190s and it is protected by a deep ditch below the terraces and visible in the photograph. Yet this court was never walled and even in 1295 had only a wooden fence around it. In 1344 it had ponds within it, perhaps those recorded in the nineteenth century. The most likely interpretation is that this court was originally the bailey of Roger Bigod's castle which was turned into a medieval garden, overlooking the mere and deer park and itself viewed from the private apartments and the castle walls above. The Bigods apparently also created a designed landscape at their nearby park of Kelsale where the remains of another lake survive.

R. Allen Brown, *Castles from the Air* (Cambridge: 1989), 119–20; R. Green, *History of Framlingham* (London: 1834), frontispiece; J. Ridgard (ed.), *Medieval Framlingham Select Documents* 1270–1524, Suffolk Record Society 27 (1985), pp. 1–11, 19; *Calendar of Inquisitions Post Mortem* IV (London: 1913), pp. 292–3; neg. no. PP 95, 14 April 1955.

Chapter 4

Renaissance Gardens

The years between 1500 and 1660 were marked by a revolution in Britain. Almost every aspect of life changed. New ideas came from abroad, long-held religious beliefs were abandoned or crushed, and great political upheavals occurred. The monasteries were dissolved, ending ancient religious traditions and leading to one of the greatest changes in land ownership there has ever been. New institutions, architecture, literature, paintings, weapons and a New World all made their appearance. In this political, social and economic ferment, it is hardly surprising that gardens too changed.

Yet, as with all revolutions, this one took place within the framework of the old world. The beliefs of the new church were embedded in the ideas and rituals of the past. The new architecture, although exuberantly exotic, was at least at first merely attached to structures from an older way of living. The nation state emerged, but from a struggle between Crown, lords and people which had begun 400 years before. Parks and gardens were the same. They too reflected new ideas and attitudes but their overall forms and much of the detail came from the past.

This dichotomy between old and new aspects of gardens poses problems for the garden historian which are made worse by the fact that, despite claims to the contrary, no sixteenth- or early seventeenth-century gardens survive in Britain. A few fragments exist but little more. Thus, the only remains at Montacute **[14]** of the 1590s are the pavilions and the form of the principal terrace. The original central mount and its associated layout have gone. What is visible is largely late-nineteenth-century in date. On the other hand, there are numerous examples of relict gardens of this period and these are a vital form of evidence, but even here care has to be exercised. Unless they had very short lives, such gardens were also altered before abandonment. The remains of the early-seventeenth-century garden at Stirling Castle **[15]** were even altered to an unknown extent later and thus are unreliable as evidence.

The value as well as the pitfalls of using only the written and illustrative evidence for gardens of this period can be seen in the works of historians such as Strong and Thacker.[1] These correctly show the complex structural and botanical development of gardens then and place them into their political, social and artistic context. These gardens are rightly seen as expressions of the power and prestige of the Crown and its wealthier subjects, displayed through emblems and symbolism. Their role as places for plays, masques and other entertainments, as well as for romantic dalliance, social meetings and philosophical or religious contemplation is also demonstrated.

Important as this approach is, it reveals only part of the history of these gardens. By looking at the, admittedly unattractive, physical remains of abandoned gardens, either from the air or on the ground, other aspects of their history and their owners become clear. The first, and perhaps greatest, value of these gardens is that there are so many of them. For the years between 1550 and 1600 the sites of at least one hundred abandoned gardens are now known. Yet, in contrast, as Strong points out,[2] there are only five gardens in England of the same period that have enough documentary evidence for their structures and layout to be worth discussing – and all five belonged to people at the very top of contemporary society. For the entire century and a half after 1500 over 250 abandoned gardens have already been discovered. Among these are those of the greatest in the land such as that at Holdenby, Northamptonshire, belonging to Sir Christopher Hatton, and also ones owned by obscure minor gentry such as that at Hamerton, Cambridgeshire.[3] The study of these abandoned gardens therefore covers a much wider social range than do the documented examples.

Another contribution that these 'archaeological' gardens make is that they contain structural details that are not known from elsewhere. The arrangement of the ponds at Kettleby **[16]**, perhaps in the initials of its owner, seems to be unique, as are the mounts of double truncated-pyramid form at Lyveden New Bield, Northamptonshire, the ziggurat-type terrace walk at Holdenby, Northamptonshire, the circular water arrangement at Croxton **[80]** and the water gardens at Chipping Campden, Gloucestershire.[4] Likewise, although documentary sources clearly indicate that many gardens were meant to be viewed from adjacent terraces or mounts, abandoned gardens show that these also provided external views, as at Alderton **[17]**, Quarrendon **[18]**, Tackley **[19]** and Croxton **[80]**. The physical remains, unencumbered by planting, also show the relative simplicity and even crudeness of many of the underlying structural elements. This reinforces the impression that the apparent neatness and permanence of parts of the illustrated gardens of the period is often illusory. At Barlings Abbey, Lincolnshire, the early-seventeenth-century garden ponds were merely monastic fishponds modified. As a result, the water arrangement there was far from an ideal neat symmetrical layout. And at Kettleby [16] the unequal knots, set within a skewed terraced layout, suggests poor design or construction.[5]

Abandoned gardens also indicate that, despite new ideas and influences, older medieval traditions continued. Although there are major differences in detail, the new Raglan gardens of the 1550s [20], set below and overlooked by the castle and themselves overlooking the adjacent deer park, have exactly the same arrangement as at Framlingham [13] of 250 years earlier or even at Stow, Lincolnshire, dating from the eleventh century.[6] Also, the form of the ponds at Alderton [17] and at Holdenby, Northamptonshire, is virtually identical to countless complexes of medieval fishponds, apart from their surrounding walkways.[7] Also interesting is the reuse of existing moats or the construction of new ones. These were made in their thousands to surround the houses of all ranks of medieval society, mainly as status symbols, and were also common in medieval gardens, as at Cawood [7].[8] At Kettleby [16] an older moat was retained and enhanced by the addition of ponds. At Hamerton, Cambridgeshire, an early-seventeenth-century moat was created as part of a water garden while at Childerley, Cambridgeshire, Quarrendon [18] and Croxton [80] the gardens were surrounded by moated features.[9]

Fewer parks were created in the sixteenth and early seventeenth centuries, the passion for hunting having lessened. Indeed, numerous old deer parks were abandoned and returned to agriculture but many survived and, particularly if they lay adjacent to great houses, remained as backcloths to new or updated gardens and as convenient hunting areas. Kettleby [16] retained its park, as did Stirling [15] and Raglan [20], while at Holdenby, Northamptonshire, a new park, set below and thus viewed in medieval fashion from the new house and garden, was created in the 1580s. It was even given plantations, including one with intersecting rides cut through it.[10]

The combination of archaeological remains and contemporary written and illustrative material show well what gardens of this period were like, even if the planting is lost. They appear to have been enclosed, bounded by high walls and hedges, and divided into compartments of varying complexity. However, the compartments could be viewed from terraces, as at Wakerley [21] and at Stainfield [79], or mounts, as at Alderton [17], as well as from contrived vistas through gates and doorways and along paths and arbours, as at

Wakerley **[21]** again. The positions of the terraces and mounts often ensured that there were also views to the outside world. These were sometimes over equally contrived parkland, but more often merely over farmland. Still-water features were common, fountains and running water less so. Knots, or flower-beds, seem to have been complex in their planting but in their surviving forms, as at Wakerley **[21]**, they appear to have been relatively simple in their basic design.

As was the case in medieval, and later, times, gardens of this period were the visual expression of the beliefs and attitudes of contemporary society. Together with their associated houses they were meant primarily to demonstrate their owners' wealth, standing and education.[11] They were intended as sumptuous settings for the adjacent houses and through their emblems and devices indicated their owners' knowledge of metaphor and allegory. They were particularly important to the new class of landed gentry who, from the spoils of the monasteries, from trade and from the law, grew rich and needed to establish their place in society. And for a few, courtiers or aspiring courtiers, gardens had an even greater importance. They were a vital requirement when visited by the monarchs who continued the medieval tradition of royal progresses. To ensure that a courtier's position was maintained or improved, not only houses and gardens but sometimes whole landscapes had to be constructed. The most spectacular was probably that at Elvetham, Hampshire, where a lake, together with islands containing emblematic structures, was created for a royal visit in 1591.[12] Another was at Kenilworth **[10]** where the medieval mere was reused in a lavish programme of events by the Earl of Leicester in 1575. The gardens at Kettleby **[16]**, Alderton **[17]** and Quarrendon **[18]** were all used, if not created, for royal visits. It was not until the reign of Charles I that the tradition of royal progresses declined.

The role of gardens as status symbols and backgrounds for entertainments makes it difficult to identify with certainty those which were made purely for the pleasure of gardening or for private enjoyment, although there can be no doubt that they existed. There was certainly much interest in plants and planting at this time. Numerous new plants were introduced and there was also an increasing number of books on gardening and garden design, but rarely can a passion for gardening be detected. One of the earliest writers was Thomas Hill (?–1576) who gave much advice on daily maintenance in what were, in effect, gardening manuals, but he rarely expressed any pleasure in gardening. Sir Francis Bacon (1561–1625) seems similar in his writings, although he certainly appreciated the aesthetics of gardens. Although eminent garden owners are recorded as having acquired specific plants, it is not at all clear to what extent these owners were really interested or whether it was fashion, or perhaps the working gardeners behind them, that were the driving force.[13] Just occasionally there are gardens that seem to have been made primarily for pleasure. One such, perhaps, is Tackley [19] for, although its owner, John Harborne, a former London merchant, is unlikely to have been responsible for the sophisticated design, its importance as a place where Harborne could indulge his passions for fishing and wildfowling seems clear.

14 Montacute House, Somerset

Although at first sight this is an Elizabethan garden, only part of the original framework survives and the remainder is restoration or idealised recreation. In the mid-sixteenth century the medieval house here was leased to Thomas Phelips, a member of an old Somerset family, once of some status but by then in decline. Just before his death in 1589 he passed Montacute to his youngest son Edward (?–1614) who increased the family's standing. Edward Phelips was a lawyer and soon became very rich. Between 1590 and 1601 he built a new house at Montacute which still stands, tall, with symmetrical elevations and all in the honey-coloured stone from nearby Ham Hill. The estate remained leased until 1608 but the house and its garden were ready for Edward's meteoric rise. He was already an MP but he was knighted in 1603, became Speaker of the Commons in 1604 and led the prosecution of Guy Fawkes. He was later Master of the Rolls.

The view here is from the east, looking at the original front entrance to the house. It was reached through the outer walled court via a gatehouse now replaced by a simple gate. The courtyard walls, together with the two corner pavilions, are contemporary with the house. The marks on the interior lawn relate to its use as a formal garden in the nineteenth century. On the north side, right, of the house and courtyard is a long terrace, also of late-sixteenth-century date, from which steps lead down into the large sunken North Garden bounded by raised grassy walks and with axial paths and a central pond. Only the external walls are sixteenth-century. The rest is a recreation of the 1890s designed by R. S. Balfour (?–1915), mainly for the last of the Phelips, William Robert (1846–1919). Originally there was a central circular mount from which the presumed knots could be viewed and certainly by 1667 there was a banqueting house on the eastern terraced walk. The rectangular grassed area bounded by trees in the right foreground, below the garden, was a bowling green with an orchard to its right. These original gardens survived until 1785 when the main entrance of the house was relocated on the west side and general landscaping took place. The orangery at the far end of the main terrace is from about 1700.

J. Brown, *The Art and Architecture of English Gardens* (New York: 1989), pp. 124–5, 128–9; A. Oswald, 'Montacute revisited', *Country Life* 118 (1955), pp. 850–3; VCH, *Somerset* II (London: 1911), p. 567; III (Oxford: 1974), pp. 212–13; neg. no. FK 45, 5 August 1950.

15 Stirling Castle, Stirlingshire

The photograph, taken from the north-west, shows the remains of what purports to be an early-seventeenth-century formal garden intended to be viewed from the great royal castle which lies immediately to the north, left. The garden lies within a much older deer park, also overlooked by the castle, and probably dating from the late twelfth century.

By the early sixteenth century, the corner of the park immediately below the castle had been set aside for a garden. It is first mentioned in 1502 and then called 'new' presumably to distinguish it from an older garden within the castle. In 1628–9 this garden was completely redesigned, a process then described as 'plotting and contryveing'. The work may have been carried out by William Watts, 'a skilful and well experimented gardener', who was brought from England in 1625 to supervise all the royal gardens in Scotland. The later history of the garden is unknown although it seems to have fallen into disuse in the eighteenth century.

The remains comprise two parts. To the north-west, in the foreground, is a complicated layout of L-shaped beds, surrounding a central circular arrangement set within a square and intersected by paths. Adjacent, to the south-east, is a similar feature but with a larger centrepiece three metres high. This consists of a double terrace of octagonal form and an inner mount set in a ditch. No exact parallels for this layout are known although its general form would fit well with the early-seventeenth-century date and later map evidence suggests that something similar to the surviving earthworks did indeed exist.

However, the site was thoroughly 'restored' in 1867 when certainly the mount was much altered. It is possible that even the orientation of the entire layout was changed then. The crisp nature of the remains reinforces this suggestion and it is doubtful if the site can be considered a genuine seventeenth-century one.

RCAHM (Scotland), *Stirlingshire* I (Edinburgh: 1963), pp. 187, 219; neg. no. BUR 87, 19 July 1975.

16 Kettleby, Lincolnshire

Kettleby lies five kilometres east of Brigg. The manor of Kettleby was acquired at the end of the fourteenth century by the Tyrwhitts, a family from Northumberland. The village was removed and its land enclosed for sheep shortly afterwards. Traces of this village, its streets, houses and the ridge-and-furrow of its fields are visible in the top left of the picture, taken from the north-east.

The Tyrwhitts soon made a name for themselves, successive members of the family being MPs, High Sheriffs or holders of Crown offices. Sir Robert Tyrwhitt (1482–1548) was a Dissolution Commissioner and as a result acquired extensive lands. Henry VIII stayed here in 1541 and there is a tradition that Sir Robert hung the trees on the road from Kettleby to Brigg with sheep and oxen to show his ability to feast all comers. Another Sir Robert (?–1617) rebuilt the house, perhaps for James I who hunted in the adjacent park. The house was abandoned in 1648 and pulled down in 1696.

In the photograph the modern farmhouse stands in the position of the old house and almost in the centre of a moated site which is probably medieval in origin. The south-east interior of the moat, left of the farmhouse, has the remains of a formal garden compartment of late-sixteenth- or early-seventeenth-century form and is thus presumably the work of the second Sir Robert. It is square, bounded by terrace walks and has traces of interior paths and flower-beds. In the left-hand, south-eastern, corner is a low circular mound, presumably the site of a summer-house. A matching mound in the north-east corner is hidden by trees. In the bottom, right-hand, north-east part of the moat is another square compartment, also with a terrace walk.

In the foreground, east of the moat and overlooked from the garden inside it, is a set of curiously arranged ponds, much damaged by modern activity, but possibly forming the letters RT in plan. This use of the initials of the builder is a typical late-Elizabethan conceit but its execution in water is unusual. The deep curving ditch in front of the ponds is the western boundary of the adjacent deer park. Beyond the moat, to the west, a series of ditched enclosures are visible. Most are probably medieval paddocks, but the large one of playing-card shape, above and south-west of the moat, is also probably part of the sixteenth-century formal arrangements.

P. L. Everson, C. C. Taylor and C. J. Dunn, *Change and Continuity* (London: 1991), pp. 70–1; neg. no. UB 10, 28 March 1957.

17 Alderton, Northamptonshire

The village of Alderton lies five kilometres south-east of Towcester. The manor was acquired, probably around 1580, by William Gorge about whom almost nothing is known. In 1582 Gorge built 'a very large' manor house on the site of the medieval one and laid out an elaborate garden. Gorge died in 1589 and his estate passed to his daughter, by then the wife of Thomas Haselrigg (c. 1563–1629) of Nouseley, Leicestershire. Haselrigg was a member of an ancient, though minor, landed Leicestershire family. In 1608, during a royal progress of Northamptonshire, he entertained James I at Alderton and was knighted there. After his death the later members of the family had little use for what they saw as the remote Alderton house and it was abandoned and then pulled down in 1700, except for a gatehouse that survived a little longer.

In this view, from the north-west, the well-preserved remains of Gorge's garden are visible. The triple courtyards of the farm buildings to the right of the photograph stand on the site of the house which faced south-west, right, towards the road. The gardens were fitted, somewhat uncomfortably, into existing closes. Behind the house site, left, is a long rectangular compartment, set into rising ground. Its right-hand third has three north-west facing terraces, while the centre part has a circular flat-topped prospect mound 22 metres across and surrounded by a broad, once water-filled, ditch. The upper third, bounded on its left by an old hollow-way leading to the village, has a slightly uneven surface, perhaps indicating that it was once a wilderness. To the north-west of the house site, in the centre foreground, is an elongated triangle of land. At its south-west end is a raised terrace overlooking a rectangular pond. There are two other ponds nearer the camera and all are surrounded by flat-topped walkways.

J. Nichols, *History and Antiquities of Leicestershire* II (London: 1798), p. 743; RCHME, *Northamptonshire* IV (London: 1982), pp. 62–3; neg. no. AWN 60, 20 January 1969.

18 *Quarrendon, Buckinghamshire*

This sixteenth-century garden lies 2.5 kilometres north-west of Aylesbury. It is surrounded by the remains of the deserted village of Quarrendon, part of which is visible in the background of this photograph, taken from the south-east.

The village was cleared in the fifteenth century, perhaps to run sheep, and probably by the Lee family who had been Crown tenants there since the late fourteenth century. The Lees lived in the old manor house which, by the sixteenth century, stood alone except for the adjacent parish church. It was probably situated below the right-hand end of the ditched enclosure. The family rose steadily in status and influence. Sir Robert Lee (?–1539) was granted the manor in 1512 and was Sheriff of Buckinghamshire in 1534. His grandson, Sir Henry Lee (1530–1610), was probably the creator of the garden here. He was one of the new Tudor landowners, a successful sheep farmer and a great encloser of land. He was also a courtier, entering royal service in 1546. He was knighted in 1553, was the Queen's Champion from 1559–90 and from 1590 was Master of the Ordnance. He was Ranger of Wychwood Forest and spent much of his time at Ditchley Park, Oxfordshire, which he bought and where he built a new house. In August 1592 Queen Elizabeth visited Quarrendon and among the entertainments for her was a masque entitled *The Message of the Damsell of the Queene of the Fayries*. Lee died childless and was buried at Quarrendon. His estates passed to a cousin and thence to three short-lived heirs. The house was abandoned probably in the 1660s and was pulled down in the early eighteenth century.

The seemingly modest garden consists of a rectangular area bounded on three sides by a formerly water-filled ditch or moat, beyond which is a raised flat-topped terrace walk. An outer ditch was also presumably water-filled. The interior was probably laid out with elaborate knots, slight traces of which are just visible. The diagonal ditch lined with trees is a much later feature. The garden may have extended further left, south-west, where the two embanked former ponds and some planting ridges might be part of it.

M. W. Beresford and J. K. S. St Joseph, *Medieval England: an Aerial Survey* (2nd edn, Cambridge: 1979), pp. 56–7, 123–4; E. K. Chambers, *Sir Henry Lee* (Oxford: 1938); VCH, *Buckinghamshire* IV (London: 1927), p. 100; neg. no. AMO 53, 1 December 1965.

19 Tackley, Oxfordshire

This unusual, if not unique, garden is here viewed from the south-west. It lies across the valley of a small south-east flowing brook nine kilometres west of Bicester and comprises a curious arrangement of one square and two triangular ponds, each with a central island of the same shape and each reached by a narrow isthmus. Around the ponds are grassy walks and there is a raised terrace on the far, north-east, side of the square pond. The whole complex lies some distance from the manor house to which it belonged and was reached by a long walk, lined by narrow canals, which extends south-west, in the foreground, below the ponds. This walk terminates at a stone archway.

The garden was created in about 1620 by John Harborne (1582–1651), a rich London merchant, who bought the manors of Tackley in 1613 and rapidly became part of the county establishment. He purchased land, undertook agricultural improvements, rebuilt the manor house and enhanced the parish church. By 1632 he was High Sheriff of Oxfordshire. He was succeeded by his two dissolute sons, one after the other, and the estate was finally leased out in 1660.

The garden as it stands is clearly incomplete, for there is a missing square pond in the top right, north-eastern, corner. A contemporary book on agriculture, Gervase Markham, *Cheape and Good Husbandry* (1623), had a plan of the complete garden inserted into it by the publisher who was a close friend of Harborne. The reason why the garden was not finished is that Harborne was unable to purchase the land on which the missing pond was to be sited. This belonged to St John's College, Oxford, who, apparently, would not sell it. The space intended for it contains two small and much older ponds, once part of an adjacent property and now hidden by the trees. Although obviously a garden, Tackley was also designed to cater for Harborne's two great loves, fishing and wildfowling.

E. Whittle and C. C. Taylor, 'The early seventeenth-century garden of Tackley, Oxfordshire', *Garden History* 22.1 (1994), pp. 37–63; neg. no. CLN 19, 18 April 1980.

20 Raglan Castle, Monmouthshire

Raglan originated in the late eleventh century but was given its great tower, walls and courts in the early fifteenth century by Sir William Ap Thomas and then much improved by his son William Herbert. Herbert, later earl of Pembroke, was executed after his defeat at the Battle of Edgecote in 1469 and Raglan eventually passed to his granddaughter Elizabeth. In 1492 she married Sir Charles Somerset (1460–1526), illegimate son of Henry Beaufort and later first earl of Worcester. The earls of Worcester were all prominent at Court and when the third earl, William Somerset (?1527–89) succeeded in 1549, he remodelled Raglan into a magnificent Renaissance palace and created gardens to go with it.

In this view, from the north, most of these gardens are visible. The most prominent features are the long terraces, cut into the slope below the castle, each with a central projection for viewing. The discovery of some stone balustrading suggests that these terraces were Italianate, perhaps with the 'curious knots' mentioned in a 1587 poem about Raglan. Below the terraces a small stream was dammed to create a sinuous lake, now drained, the bed of which can be identified by the modern drainage ditches in the valley bottom, visible in the foreground. This lake, the terraces and a deer park which extended north beyond the lake could all be viewed from the Long Gallery which the earl added to the castle above. The gardens also extended to the south-west, right, of the castle, with further terraces, a raised bowling green and an orchard. When the fourth earl, Edward (1563–1628), succeeded in 1589 he enlarged the gardens. His work included a rectangular 'water-parterre' at the head of the lake with triangular and diamond-shaped islands and a central feature. This survives in outline, visible at the bottom left of the picture. The earl also created another water garden which was square with four square islands in the valley to the south.

The fifth earl supported Charles I and was made Marquis of Worcester. In 1646 the castle was besieged, captured and rendered indefensible. When Henry Somerset, the grandson of the first marquis, required a new home in 1673, times had changed and he created a new house at Badminton, Gloucestershire. Raglan was then abandoned.

E. Whittle, 'The Renaissance gardens of Raglan Castle', *Garden History* 17.1 (1989), pp. 83–94; neg. no. AXA 3, 3 February 1969.

21 Wakerley, Northamptonshire

The remains of this magnificent garden are at Wakerley, 12 kilometres north-east of Corby. Its exact date is unknown but it is likely that it was created in the 1620s. The site was occupied by a farmhouse in 1615, when it was leased to a local man, then sold in 1618 to Sir William Cecil (c.1567–1633), second son of the first earl of Exeter, who again leased it. However, when Cecil died he was described as 'of Wakerley' and the garden, together with its associated house, by then existed. Part of the house continued to be occupied by Cecil's daughter but as his son became the third earl the family eventually had no use for it. It was therefore pulled down and the garden abandoned in about 1690. By 1715 only a barn remained.

In this view from the north, the uneven foundations of the house lie to the west, right. Above it are the robbed-out wall footings of at least five rectangular compartments. The two compartments immediately south of the house have traces of internal arrangements as well as a simple axial path extending from the house. To the east, left, of the house is a large rectangular area bounded on the south and east by double terraces. Its interior has the slight remains of an axial path which crosses a complex octagonal layout of beds and paths in the centre and groups of low ridges at the east end. The foundations of the large building, above and to the right of the house site and partly cut by the road of 1749, are presumably those of the eighteenth-century barn.

RCHME, *Northamptonshire* I (London: 1975), p. 105; neg. no. BLG 93, 15 December 1972.

Chapter 5

The Age of Geometry

From 1660 to about 1730 the parks and gardens of Britain continued to be altered or created anew in response to changes in political, social and aesthetic attitudes. Continental influences on British garden design, especially Italian, were important even in the sixteenth century and were widely accepted particularly by the wealthy and influential,[1] but after the Restoration in 1660 these continental influences became a flood. The reasons for this are various. One was that royalists, exiles during the Commonwealth, returned to reclaim and refurbish their estates, having seen and lived with many of the greatest European gardens. Another reason was the impact of the court of Louis XIV which was so magnificent and so outshone anything else in Europe that many of its visible trappings, including garden designs, were eagerly adopted by European upper classes. In addition, the gardens of Holland, themselves derived from Italian and French models, became fashionable after 1689 when England accepted a Dutch king, William III. The long-established French connection with Scotland, exemplified by the support for the Jacobite succession, was also manifested in many gardens there.

Also increasingly important was the Grand Tour which no self-respecting, educated gentleman could fail to make. These tours meant that the participants saw the best of contemporary gardens and that they returned home with real looted antiquities, or suitable imitations, to display in gardens which they now knew how to create. The most important of these European influences were Italian but they emanated not so much from contemporary gardens as from those associated with the great Renaissance palaces. These gardens were strongly architectural with the adjacent house often projected into the garden which became another room or rooms. These Italian models were usually on sloping ground and were thus terraced, linked by steps and adorned with statuary. The planting was ordered and formal and contained within compartments but, because of the situation, both inward and outward views of the countryside beyond were possible. Water from grottos, fountains and jets was fashionable as were cascades and pools. Outer, semi-formal groves, cut by walks and punctuated by statuary were common.[2] Some of these details, albeit modified, may be seen at Powis [22] with its gigantic late-seventeenth-century architectural terraces, extending the house into the garden, as well as its wide views. Powis was apparently inspired by a French garden in an Italianate style which its exiled owner knew. The almost contemporary but less impressive terraces at Harrington [23], which also give views across an agricultural landscape, are again Italianate in conception. So too are the terraces of 1712 at Gamlingay [24] which overlooked a trapezoidal lake and formally planted woodland. The garden at Gamlingay was also the result of a Grand Tour by its owner, as was the accompanying house.

Late-seventeenth-century French gardens had an even greater impact in Britain. Although Italian in inspiration, these French gardens had somewhat different characteristics. These included often extensive parterres, with complex curvilinear planting, below the house. Terraces were also a feature and topiary cut into architectural shapes was fashionable. Complicated arrangements of ponds, geometric lakes and fountains were also common. Beyond the gardens the surrounding areas, with straight avenues, formal plantations and rides, were an integral part of the layout thus demonstrating the taming of Nature.[3] The remains of a post-Restoration royal garden, in an almost entirely French style although perhaps unfinished, survive at Greenwich Park, London. The terraces there, as well as the framework of the surrounding formal parkland, still exist.[4] However, the best surviving example of a late-seventeenth-century French-inspired garden and park in Britain, although also largely an archaeological site, is that at Boughton [25], created between 1683 and 1709 by the former English ambassador to the court of Louis XIV.

Elsewhere the terraces, canal and large parterres of the 1660s at Wing [26], the parterres of 1712 at Gamlingay [24] and of 1722 at Eastbury [78], the topiary of 1700 at Levens [27], the avenues of 1707–11 at Heythrop [28] and of the early eighteenth century at Tredegar [74], the formal woodland of c. 1710 at Wrest [29] and almost the entire landscape of 1700–25 at Bramham [30] seem, in various ways, to be based on French designs. Wing was created by its owner on return from exile, Gamlingay, Wrest and Bramham after Grand Tours, and Levens was by a French gardener.

Gardens of this period in The Netherlands derived largely from France but were somewhat different in that parterres were more complex and canals, especially flanking canals, more common.[5] In Britain the small garden of 1696–1703, with its long canal, at Westbury, Gloucestershire, is the only surviving complete if restored example.[6] And although the Dutch parterres at Wrest [29] have gone, the main canal remains, as do the much modified flanking canals. Perhaps the best known Dutch-inspired garden is the Privy Garden at Hampton Court, again a restoration, but based on detailed excavation and documentary research.[7]

The major British gardens of this period, however, were designed to express political events and ideas. Especially in the early eighteenth century these ideas, when combined with new philosophies and aesthetics, produced different gardens and parks. Initially these political and philosophical ideas were concerned with architecture whereby a more 'correct' use of classical designs became the hallmarks of Whig politicians, after their victory in 1714 on the accession of George I. This was in contrast to the less rigorous baroque style said to represent both the earlier Tory governments and the decadent monarchies of the continent. The resulting new climate created what were judged to be ideal houses, such as Chiswick and Mereworth, but also led to more extensive developments in parks and gardens. This manifested itself in the appearance of classical temples and statuary but, more significantly, in an increasing simplification of the previous complex layouts, with cleaner lines in both planting and associated buildings and an increasing use of grass. The rigid geometry remained but gradually what have been termed 'stripped-down' landscapes appeared, characterised by the work of Stephen Switzer (1682–1745) and Charles Bridgeman (?–1738).[8]

The greatest of these 'stripped-down' landscapes is perhaps Studley Royal [31] where, between 1722 and 1742, its owner and creator took a French design and turned it into an English landscape by making its parterres of water and then setting them in an informal wooded valley. Another, probably British, invention which also allowed the development of different landscapes was the ha-ha or sunken fence which enabled views of gardens to be extended easily into the adjacent parkland. Ha-has appeared in the late seventeenth century, as at Levens [27] in 1694, and were much used by Bridgeman and his successors.[9] Parks, however, remained as formal as before as, for example, at Bridgeman's own creation, Eastbury [78].

To what extent these practical, political, artistic and philosophical ideas were fully understood by the owners of all the new houses, parks and gardens of this period is by no means clear. Fashion and a fear of being different were always powerful factors in the creation of gardens and this was especially true of the early eighteenth century when 'culture' was being advocated as a new concept. This is not to say that philosophers, writers, artists, architects and garden designers did not genuinely believe in what they advocated. They certainly did believe in it, and there were serious and almost endless debates on 'taste'. There was a nostalgia for an imagined perfect past and a desire to re-establish a stable world, but there was also a much wider 'culture' which people accepted and imitated even if it was not fully understood. Many landowners, of both the aristocracy and the gentry, looked to poets, artists and philosophers to tell them what to read, what to believe, what to admire and finally what kinds of parks and gardens to have. And, in the end, these parks and gardens could be sold to them 'by the yard' in the name of taste.[10]

Complex as all these developments were, there is yet one more aspect of parks and gardens of this period to be considered. For, as always, in addition to the ferment of new ideas and designs, elements of traditionalism and conservatism also existed. Despite the changes all around them, many owners for a variety of reasons left their parks and gardens untouched. Indeed, even in the late eighteenth century there were numerous gardens which were still virtually as they had been 150 years earlier.[11] Although garden historians inevitably, and perhaps correctly, concentrate on new developments, these unaltered gardens actually formed a very large proportion of the total in existence in the eighteenth century. An example is Bradgate [32], the gardens of which were somewhat old-fashioned even when they were created, probably in the 1660s. Despite this they were never altered, perhaps because their owners disliked change. Another instance is Gamlingay [24] which, although most of the garden there was in the height of fashion when it was laid out in 1712, also included a group of walled compartments that could have been designed a century earlier. Even those gardens displaying continental influences may not be entirely without some traditional input. The terraces at Powis [22] may have had a French prototype but similar arrangements may be seen at a much earlier date at Raglan [20] and indeed some of the basic forms of early-eighteenth-century parks and gardens already existed in medieval times.

Inevitably, the perceived reasons behind the creation of most gardens and parks of this period seem to be fashion, status, political correctness and above all 'taste'. There is little obvious evidence of dedicated gardeners. With gardens such as Eastbury [78], which seems to have been little more than a vastly expensive stage for an eccentric's flamboyant social life, it is difficult to detect a real passion for the gardens themselves. Yet, there was also an underlying interest in, and a deep love of, plants and planting throughout the period. New introductions, increasingly from the New World, were avidly sought after, writers such John Evelyn (1620–1705) were widely read, and nurseries and nurserymen flourished.[12] And although uncertain, sometimes the person behind the gardens emerges. Most notable perhaps is John Aislabie who spent twenty years after his enforced retirement from public life creating what is arguably the greatest garden in Britain, Studley Royal [31]. Levens [27] too, on a much smaller scale and set around the family home of another political exile, may also be a gardener's garden. Wrest [29] and Bramham [30] also seem to be further examples of the close involvement of owners. Less certain is the landscape at Ashdown [33] which seems to have been created, together with its tiny house, as a hunting park and lodge, purely for pleasure and remote from its owner's larger house and estate. Although up-to-date in appearance, the ideas behind it can be traced back through sixteenth- and seventeenth-century lodges and their landscapes, such as Wothorpe, Cambridgeshire, Lyveden, Northamptonshire, and elsewhere, to Framlingham [13] and Harringworth, Northamptonshire, in medieval times.[13]

22 Powis Castle, Montgomeryshire

In 1587 Sir Edward Herbert purchased the medieval castle of Powis and began its transformation into a country house. The Herberts were staunch royalists and Catholics. They served all the Stuart kings with great loyalty and were rewarded by being elevated through the ranks of the peerage. William, the third baron Herbert (1626–96), was created Earl of Powis in 1674 and Marquis of Montgomery in 1686. In 1688 he followed his king into exile and he, and possibly his son William, lived at St Germains-en-Laye, near Paris. It may have been the magnificent Italianate terraced gardens there that influenced the later work of the second William (c. 1665–1745) at Powis. However, it is likely that the first marquis had already begun the gardens at Powis in the early 1680s, perhaps under the direction of the architect William Winde (c. 1640–1722) who was then engaged in altering the castle.

During the Herbert exile the castle and estates were granted by King William to his cousin, the Earl of Rochford, but he never lived at Powis. Although technically the Herberts did not regain Powis until 1722, William the second marquis returned home in 1703, bringing with him his wife's French gardener Adrian Duvall. It was probably soon afterwards that Duvall, under the supervision of Winde, completed the gardens.

The view here is from the south, across the four great 200-metre long terraces which are said to have been blasted out of the hillside. Although the terraces extend well beyond the house, their architectural components are confined to the 70-metre long area immediately below it, so giving an appearance of great height and grandeur. Since the photograph was taken they have all been replanted by the National Trust. The upper terrace has a series of elaborate niches and is now topped by bushy yews much larger than they were originally. The next terrace has eight arched brick openings leading into what was once an aviary. The centre of the third terrace is wider and is backed by an orangery with large windows. The lowest terrace is plain. The grassy valley floor below was originally laid out with a formal Dutch-style water garden with pools and fountains. This was removed in the late eighteenth century.

R. Haslam, *The Buildings of Wales* (London: 1979) pp. 188–96; C. Hussey, *English Gardens and Landscapes* (London: 1967), pp. 53–6; National Trust, *Powis Castle Gardens* (London: 1992); E. Whittle, *The Historic Gardens of Wales* (London: 1992), pp. 30–1; neg. no. EC 7, 22 July 1949.

23 *Harrington, Northamptonshire*

The manor of Harrington, 10 kilometres west of Kettering, was purchased in 1599 by Sir John Stanhope (?–1621) who enlarged the medieval house there. Stanhope was created Baron Stanhope of Harrington in 1605. He was succeeded by his son Charles who lived at Harrington until the Civil War when he was exiled to France. On his return he moved to Lincolnshire and died childless in 1675. Harrington then passed to Elizabeth, Countess of Dysart, who had married William Murray and was the daughter of Charles Stanhope's sister. Murray, a clergyman from Dysart in Fife, was secretary and tutor to Charles I when Prince of Wales. He remained close to the King and was created Earl of Dysart in 1643. In 1647 the Countess married Sir Lionel Tollemache of Helmingham, Suffolk, and on her husband's death in 1653 obtained licence to nominate her descendants as earls of Dysart. Her son Lionel (1648–1727) was thus the second earl and it was he who acquired Harrington on his great-uncle's death in 1675 and who further improved the house and created the gardens. These were certainly complete by 1712 but their later history is unknown. The house was probably abandoned by 1750 and had disappeared by the early nineteenth century.

The gardens here are viewed from the north-east. They lie on land sloping gently north-west, left to right. The house stood at the lowest end and its site is marked by the uneven ground in the middle distance, right. It was approached by a drive running along the north-east boundary of the garden, with a line of trees on its outside edge. Some of the trees survive in the left foreground and others are marked by stumps and holes. Ring-counts indicate a planting date of about 1700.

Behind the house site is a rectangular sunken compartment with a central circular depression, once a pond, and raised paths, bounded on three sides by double terraces. To the left, the rising ground is cut into five terraces each with a different arrangement. The lowest has a recessed centre with steps in the middle and sunken paths. The next also has traces of paths while the middle one has a large trapezoidal former pond in the middle. The next terrace up also has traces of paths and two square former ponds while the top one had a diamond-shaped pond in the centre. The ovoid trench-like feature extending across all the terraces is a continuous walkway which allowed complete access around the garden.

J. Bridges, *History of Northamptonshire* II (London: 1791), p. 33; J. Morton, *Natural History of Northamptonshire* (London: 1712), p. 494; RCHME, *Northamptonshire* II (London: 1979), pp. 75–7; neg. no. SB 58, 28 March 1956.

24 Gamlingay Park, Cambridgeshire

Gamlingay lies in south-west Cambridgeshire, four kilometres north-east of Sandy. The estate, including a house and park, was bought by Sir George Downing I soon after 1660. He had been a leading member of the colony of Massachusetts but returned to live in Suffolk in the 1640s while amassing numerous properties elsewhere.

In 1700 his grandson, Sir George Downing III (1685–1749), when only fifteen, secretly married his thirteen-year-old cousin. Two years later he left for a Grand Tour and on his return attempted to repudiate the marriage. After ten years of dispute this was rejected by the House of Lords. In 1712–13 Sir George, presumably influenced in part by his continental journeys, built a new house at Gamlingay and surrounded it with an elaborate garden. In 1717, in an attempt to prevent his estranged wife inheriting his land, he devised a complex plan involving four cousins and the establishment of a Cambridge college. Except for the eventual founding of Downing College in 1800 the plan failed. The house was demolished and the garden abandoned in 1776.

The view here is from the south. The house, in a sixteenth-century Italianate style, consisted of a central block with south-projecting wings. It lay just above and on either side of a sunken lawn, the circular feature in the foreground. The linear depression beyond marks its cellars. To the right of this lawn and the site of the east wing, irregular ground marks the position of outbuildings. To the left, the narrow lines are the robbed-out footings of a group of former walled compartments. Immediately above, north, of the house site is a long terrace with ramps at each end. These lead down to a large rectangular level area with traces of a central path. North again, further ramps descend to a lower, crescent-shaped terrace. Beyond is the bed of a trapezoidal lake. Its northern and western edges are marked by hedges and its eastern side by a 250-metre long dam, four metres high. Water from this lake flowed south below the dam and fed a line of rectangular ponds, the remains of which are hidden by trees. Originally these ponds could be viewed from the walkway along the dam or from the terraces above. The land to the north of the lake was formerly almost completely wooded with numerous intersecting straight rides.

RCHME, *West Cambridgeshire* (London: 1968), pp. 110–12; VCH, *Cambridgeshire* III (Oxford: 1959), p. 487; VCH, *Cambridgeshire* V (Oxford: 1973), p. 74; neg. no. ARH 83, 16 May 1967.

25 Boughton House, Northamptonshire

In 1528 Boughton, four kilometres north-east of Kettering, was purchased by Edward Montagu (c. 1486–1557) a successful lawyer and later Chief Justice. With his increasing wealth he bought more land so that, on his death, he owned a large compact estate, centred on Boughton, where he also enlarged an earlier deer park. Montagu's grandson Ralph (1638–1709) was a politician and holder of many Crown appointments, including that of ambassador to France in 1669–72 and 1676–8. Although his fortunes declined in the 1680s, he supported the new Protestant king and was made an earl in 1689 and a duke in 1705. On inheriting Boughton in 1684 he began a programme of house-building and garden creation that was still unfinished at his death. The house has been described as 'more French than any other contemporary house in England' and a Dutch gardener, Leonard van Meulen (fl. 1685–1710), spent over twenty years laying out an equally French garden. The archaeological remains of this garden survive and are the best of their date in Britain.

In this view, from the west, Boughton House lies in the far distance. In front of it, now edged by multiple rows of trees, was a great parterre, the boundary terraces and former flower-beds of which remain. Below it, at the far end of the long avenue is the site of a rectangular lake. To the left, north, of the parterre, between the trees, are the remains of a long canal and a square pond edged by terraces. To the right of the parterre is another smaller parterre, a water garden, a cascade and a formal wilderness. The latter is the rectangular plantation on the extreme right. The western boundary of the garden was the small River Ise which was turned into a series of embanked straight canals, three of which lie beneath the Z-shaped line of trees, centre right.

John, the second duke (1688–1749), simplified the great parterre in the early 1720s, and in the late 1720s and 1730s he further simplified the layout and planted lines of trees along the canals and terraces. He also began the creation of a system of avenues, plantations and rides through existing woodland which, by the 1750s, included some 36 kilometres of avenues spread over 50 square kilometres and most of eight parishes. Part of the principal western double avenue is visible in the foreground and some of the new woodland is in the distant background. On the duke's death, Boughton passed by marriage to the dukes of Buccleuch. This family neither used Boughton much nor greatly altered its surroundings which thus survive close to their early-eighteenth-century form.

RCHME, *Northamptonshire* II (London: 1979), pp. 154–62; J. Heward and R. Taylor, *Country Houses of Northamptonshire* (London: 1996), pp. 94–109; neg. no. EZ 71, 10 July 1950.

26 Wing, Buckinghamshire

A former monastic estate at Wing was acquired in the sixteenth century by a member of the Dormer family. The family wealth derived from a late-fifteenth-century Oxfordshire wool merchant. In the mid-sixteenth century Sir Robert Dormer built a house here and probably created the adjacent parkland. The family were staunch royalists and Sir Robert's son was created first Baron Dormer in 1615. His grandson, another Robert, was created Viscount Ascott when only eighteen and first earl of Carnavon in 1628. He was killed, fighting for his king, at the first battle of Newbury in 1643. It may have been the first Earl who, in addition to enlarging the house, created the magnificent gardens whose outlines can be seen in this view from the south-east. However, the house was ransacked after the earl's death and later restored by his son, the second earl, Charles Dormer (1633–1709), who lived at Wing in great style and entertained lavishly. The garden is thus more likely to be his work and, stylistically, a date after 1660 would be acceptable.

The location of the house, about which nothing is known, is marked by the uneven ground in the background, right. A terrace in front of it overlooks a rectangular area with a central former pond, outer pathways and raised side walks. This is succeeded in turn by a double terrace overlooking another rectangular area, also with raised side terraces, the interior of which shows traces of paths and parterres. Below, and beyond a linear depression which presumably was once a canal, is yet another rectangular space crossed by the modern track. The uneven surface here suggests that this was once wooded. Beyond again, as the land rises, two parallel banks mark the line of a vista which terminates in what is now a ploughed-out mound covered with seventeenth-century bricks. To the left of the apparently once wooded rectangle are the shrunken remains of a formerly much larger embanked lake. Above it, in a broad straight depression, once another long canal, is its feeder stream. Left of this canal, parallel to it and extending alongside the lake are the remains of two groups of numerous crescent-shaped flower-beds set within rectangular ditched areas, possibly originally walled. These are extraordinary features with no parallels anywhere. When the second earl died without heirs the property passed to cousins who lived elsewhere and the house and gardens were abandoned.

G. Lipscombe, *History and Antiquities of Buckinghamshire* (London: 1847), pp. 524–5; VCH, *Buckinghamshire* III (London: 1925), p. 45; neg. no. NV 46, 24 April 1954.

27 Levens Hall, Westmorland

Levens lies seven kilometres south of Kendal. In 1686, the sixteenth-century house and estate were bought by Colonel James Grahame (1649–1730), MP, soldier, Jacobite, Keeper of Bagshot Park, Privy Purse to James II and close friend of the king. His intention was to make Levens a family home. He was arrested in 1691 by the new government because of his political leanings and, although freed, continued to live at Bagshot Park until 1699. The garden at Levens, laid out between 1689 and 1712, was designed by Guillaume Beaumont (fl. 1684–1727), a Frenchman who had a nursery at Bagshot and who had designed a garden at Hampton Court. He had worked under Grahame when the latter was effectively supervisor of the royal gardeners from 1683 to 1689.

The garden, seen here from the north, comprised five geometrical 'quarters', bounded by yew and beech hedges, some of which survive. The most remarkable of these quarters is the Parterre or Topiary Garden left and immediately east of the house. This is a rectangle with axial and surrounding paths separating and bounding an amazing collection of clipped bushes and trees, now underplanted with modern bedding plants within box-edging. Whether the present fantastic shapes of the topiary are close to the original forms is doubtful, although most of the trees may be seventeenth-century. They were recut in about 1815, after years of neglect. Above are the other four compartments, set around a central circular open space. These were originally an orchard, a soft-fruit area, a bowling green and a melon-ground, but now have modern planting. The main east to west axial path terminates at the west on a bastioned ha-ha, overlooking the adjacent park which still contains remnants of its late seventeenth-century avenues. The ha-ha, dating from 1694, is one of the earliest known.

Grahame made his peace with the monarchy in 1701 and returned to public life. Beaumont continued to make small alterations to the garden until his death in 1727.

A. Bagot, 'Monsieur Beaumont and Colonel Grahame, the making of a garden', *Garden History* 3.4 (1975), pp. 66–78; G. Jellicoe (ed.), *Oxford Companion to Gardens* (Oxford: 1991), p. 336; neg. no. MZ 3, 10 August 1953.

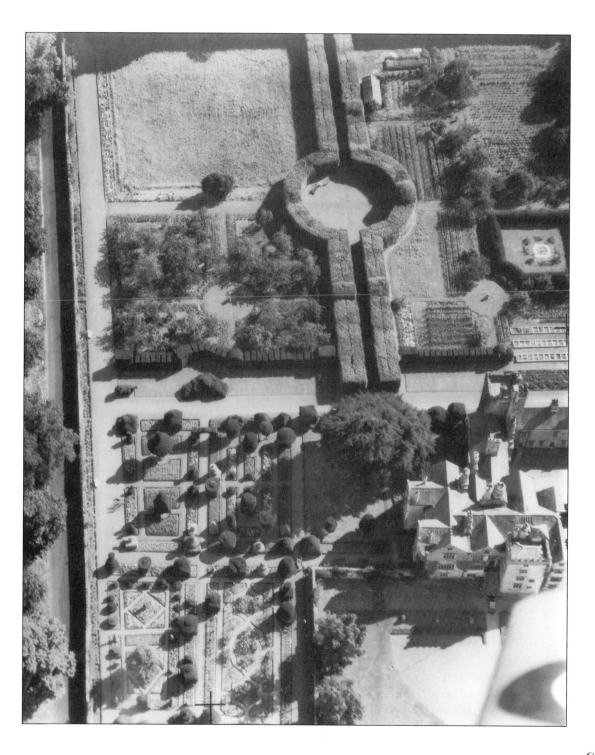

28 Heythrop Park, Oxfordshire

Heythrop lies five kilometres east of Chipping Norton. It was purchased in 1697 by Charles Talbot, twelfth earl and later first duke of Shrewsbury (1660–1718). Talbot abandoned his Catholic faith and conformed to the established church in 1679 and in 1688 was one of the seven signatories to the invitation to William III. He was a favourite at court and was Secretary of State in 1689 and in 1693–8. He was also a close friend and near neighbour of the Duke of Marlborough with whom he certainly consulted over both their new houses. As a result of his Crown offices Talbot would have known Henry Wise (1653–1738), the royal gardener, who may have been involved with the park at Heythrop when he was living and working at Blenheim between 1705 and 1716. From 1700 to 1705 Talbot toured Italy and returned home with much statuary, the plan of a Florentine palace and an Italian wife. In 1706 he engaged Thomas Archer (?1668–1743), the only English architect then to have studied in Italy, to design a new house at Heythrop. The house and garden were constructed together between 1707 and 1711.

Although the park had a wooded 'natural' wilderness to the south-west of the house, with winding paths, a rill, a bath and a cascade, most of it was very formal. The main axis, which passed through the house, was projected south–east by a short tree-lined vista and to the north–west by three-kilometre-long avenues, the vanishing point of which was the horizon. These avenues were made up of two inner rows of alternate square clumps of forest trees and circular ones of silver firs, with outer multiple lines of trees. A simpler and much shorter avenue of clumps extended at right angles north-east with another tree-lined vista to the south-west.

The view here, from the south-east, shows the much altered, incomplete and partly replanted great north-western avenues extending into the distance. The rectangular plantation in the right foreground was once a formal grove and still contains the outlines of eight walks radiating from a sunken circular bowling green.

The Talbots remained at Heythrop until 1820. The house was burned out in 1831 and was then bought in 1870 by Thomas Brassey (1805–70), the railway contractor, for his son who restored it. The park remained reasonably intact until 1918 when the estate was broken up.

'Heythrop, Oxfordshire', *Country Life* 18 (1905), pp. 270–6; R. Bisgrove, *The English Garden* (London: 1990), p. 79; N. Pevsner, *Oxfordshire* (London: 1974), pp. 646–9; VCH, *Oxfordshire* XI (Oxford: 1983), pp. 134–5; neg. no. GW 89, 19 July 1951.

29 *Wrest Park, Bedfordshire*

Wrest Park lies six kilometres south-east of Ampthill. In common with many gardens the successive phases of its history are evident. The view here is from the south-west with the house in the left background. The earlier house stood further south, just below the circular fountain.

Wrest belonged to the Grey family from the thirteenth century. In 1676 Anthony Grey (1645–1702), the eleventh earl of Kent, with his wife and mother, refronted the old house and began to transform the existing seventeenth-century garden. The section immediately below the house was given Dutch-style parterres with a path leading to a long canal edged by trees and extending across open parkland. This canal lies within the trees in the central foreground. The twelfth earl, Henry Grey (1671–1740), did the Grand Tour and then served Queen Ann as Lord Chamberlain. For this he was made a duke in 1710. However, his preoccupation was with his garden at Wrest which he developed for over thirty years. He created formal woodland areas on both sides of the canal which were cut through by straight rides intersecting at or leading to small glades. These areas and some of the rides survive. In the foreground, at the end of the canal, is the domed baroque Pavilion by Thomas Archer, built in 1709–11. The duke also introduced formal canals surrounding the wooded areas and grassed over the parterres near the house. On the west, just above the left-hand wooded area, a Bowling Green was created in 1735, together with a classical Bowling Green House by Batty Langley (1696–1751), which is just visible above the trees. Although the duke sought advice from many of the leading contemporary architects and garden designers, most of the planning was done by him with the help of his family and his gardener, John Dewell.

The duke was succeeded by his daughter Jemima (1723–97) who married Philip Yorke, son of the earl of Hardwicke. Ignoring the more fashionable informal landscapes of the late eighteenth century she changed little. Even though Capability Brown was employed he did nothing more than make the external canals less formal. Part of his work can be seen in the sinuous canal in the foreground. In 1833 Wrest was inherited by Thomas, Earl de Grey (?–1859), a francophile. In 1834-6 he demolished the old house and built a new one in a French style 200 metres to the north. In front of it he constructed an equally French parterre, as well as an orangery to the south-west. After 1859 the house was used only occasionally although the gardens were maintained. The estate was sold in 1917 after the death in action of the last direct heir. In 1946 it was purchased by the government and the garden is now in the hands of English Heritage.

T. Richardson, 'Wrest Park', *Country Life* 190 (1997), pp. 38–43; neg. no. FF 39, 14 August 1950.

30 Bramham Park, Yorkshire, West Riding

Bramham lies 15 kilometres north-east of Leeds. The site was an empty one when it, and a fortune, were left to Robert Benson (1675–1731) by his father, another Robert (?–1676), who, from lowly beginnings, became a successful lawyer. Benson went on the Grand Tour and on his return began to build a house and to lay out the landscape at Bramham. He was a self-made man, successful in business and politics and an MP and later a government minister. He resigned in 1713, was created Lord Bingley and became ambassador to Spain. He was a director of the South Sea Company 1711–15 and Treasurer of the Household 1730–1.

The house at Bramham was begun in 1699 and completed c. 1710. The park and garden, which Benson seems to have designed himself, perhaps aided by his gardener Robert Fleming, took much longer to finish. Alterations were made in the 1720s, probably by John Wood the Elder (1704–54). On Benson's death Bramham passed to his daughter Harriet who had married George Lane-Fox (?–1773), also an MP, later ambassador in Vienna and created Lord Bingley in 1763. Together they continued to improve the gardens, adding a number of structures, some by James Paine (1716–89). Lane-Fox died without direct heirs and Bramham passed to a nephew. Thereafter the gardens were neglected and a fire in 1828 led to the abandonment of the house until 1907.

The view here is from the north-west and covers only the garden, not the extensive park and woodland. The house stands in the left foreground and to its right is the sunken walled parterre, now a rose garden, cut into the rising ground with the remains of a cascade at its head. Between the house and the parterre and extending south is the Broad Walk. This is on the main axis of both the gardens and of the woodland to the south. At its north end, in the foreground and half hidden by trees, the Walk terminates at an orangery (by Paine c. 1760), later a chapel. To the south, the Walk ends at a set of formal ponds, visible near the top of the picture, left of centre. The vista continues through the woodland over a formal Great Cascade (c. 1725) into a valley. Beyond, out of the picture, the vista extends upwards across the park to an extensive pleasure grove divided by numerous intersecting straight walks. The main vista is blocked, first by an Ionic temple (by Paine 1750) and then by an obelisk (1768).

To the right, west of the Broad Walk, the wooded garden, although altered by the addition of later features such as tennis courts and by the loss of trees, retains some of its *allées* flanked by high beech hedges. Within one, upper right, is a T-shaped Canal (1728). Above and left of the Canal, towards the right-hand end of the open glade, is a Gothic Temple (c. 1750).

C. Hussey, *English Gardens and Landscapes* (London: 1967), pp. 70–7; A. Oswald, 'Bramham Park', *Country Life* 123 (1958), pp. 350–3, 1294–7, 1368–71; neg. no. CLV 53, 17 May 1980.

31 Studley Royal, Yorkshire, West Riding

In 1699 John Aislabie (1670–1742), son of a prominent lawyer, inherited from his mother's family an estate at Studley, three kilometres south-west of Ripon. Aislabie, an ambitious man, was Tory MP for Ripon and by 1718, after a convenient change of allegiance, Chancellor of the Exchequer for the Whig government. He was also a keen gardener and in 1713 was assisting his son-in-law, Edmund Waller, with the gardens at Hall Barn, Buckinghamshire.

In 1716 Aislabie began work at Studley. He was a sponsor of the South Sea Company and had promoted the parliamentary bill for it but when the South Sea Bubble burst in 1720 he was briefly imprisoned and disqualified from holding public office. He retired to Studley and spent twenty years creating in a wooded valley in a corner of his estate what was to be one of the greatest formal gardens in the world. On his death, his son William (1700–81) extended the layout in a more romantic style. The upper part of the valley, which included the remains of the medieval Fountains Abbey, was purchased in 1768. The subsequent landscaping there is an early instance of the picturesque style.

Only a small part of the garden can be seen in the photograph. Here, viewed from the north-east, is the lower part of John Aislabie's layout. In the wooded valley it is formal and geometric and clearly inspired by contemporary French gardens. Yet lawns rather than elaborate parterres separate water features which mirror the surrounding trees and buildings. In the foreground above a lake is the Cascade, flanked by classical Fishing Houses. Behind it is the Canal with the small Half-Moon Pool to its right. On the left is the Moon Pool with a statue of Neptune in its centre and the Temple of Piety (1742) hidden in the trees, on its east side. This pool is flanked by two Crescent Pools. At the head of the Canal are the circular Drum Falls, fed by another canal. On the hillside to the right an open glade contains the Banqueting House (1728–32), possibly by Colen Campbell, and the remains of a sunken garden. On the opposite hillside the upper part of the classical Octagonal Tower (1778) with its later Gothick pinnacles is visible. All this is only a quarter of the whole garden. Further upstream is another Half-Moon Pool and a canalised river extending to the Abbey ruins. The entire garden is now a World Heritage Site.

S. Beard, 'A magnificent landscape garden', *Country Life* 130 (1961), pp. 284–7; C. Hussey, *English Gardens and Landscapes* (London: 1967), pp. 132–9; National Trust, *Fountains Abbey and Studley Royal* (London: 1989); neg. no. CQD 17, 18 June 1984.

32 Bradgate Park, Leicestershire

Bradgate Park lies north-west of Leicester on the edge of Charnwood Forest. It was created as a deer park before 1247 but the house was not built until the late fifteenth or early sixteenth century either by Thomas Grey (before 1455–1501), Marquis of Dorset and stepson of Edward IV, or by his son, another Thomas (before 1478–1530). The Grey family continued living at Bradgate and it was either Henry Grey (1600–73), created Earl of Stamford in 1628, or his grandson, Thomas (1653–1719), the second earl, who improved the house and created the gardens there, probably after 1660. Both lived there in some style. Even after improvements, the house remained old-fashioned by contemporary standards, as did the gardens. Indeed, their interest lies in their archaic arrangement, despite a visit by William III in 1696.

The view here is from the east. The ruinous house of U-shaped plan stands in the left background. The gardens lay within a series of walled enclosures which survive and within which are traces of both structures and planting arrangements. These can be compared to the 1707 illustration by Kip and Knyff. The rectangular compartment below the house has remains of terraces on all four sides. The former axial and surrounding paths are visible and Kip shows plain parterres between them, perhaps lawns, as the photograph indicates. The compartment below again, in the foreground, has lost its north, right-hand, boundary but the upper terrace and long canal, all shown by Kip, survive, as does the raised central path. To the south, left, of the path, where Kip shows lines of trees, low planting ridges exist, as do traces of flower-beds to the north of the path where Kip also shows beds.

Along the north side of the house and extending east along the north side of the two compartments described above, Kip shows a single large rectangular compartment, subdivided by hedges, lines of shrubs and a fence. It had scattered trees in its eastern half, two lawns and perhaps an orchard at the west end and, in the centre, a rectangular pond with a copse on its north-west side. On the ground the area of scattered trees has planting ridges on it, perhaps indicating former lines of trees. In the centre are three ponds, one L-shaped. The copse remains as do the two lawns beyond, together with their boundary hedges, now ditches.

The widow of the second earl lived at Bradgate until her death in 1722. The estate then passed to a Staffordshire cousin who had no use for an old-fashioned house and garden. They were therefore abandoned.

W. G. Hoskins, *Leicestershire* (London: 1957), pp. 37–8; J. Kip and L. Knyff, *Britannia Illustrata* (London: 1707, reprinted 1984), plate 12; N. Pevsner, *Leicestershire and Rutland* (2nd edn, London: 1984), pp. 108–9; neg. no. AMY 85, 30 January 1966.

33 Ashdown House, Berkshire

Here are the last remnants of a mid-seventeenth-century designed landscape, the origins of which have been romanticised by an unsubstantiated association with Queen Elizabeth of Bohemia, the Winter Queen for whom, it has been alleged, the house was built. It was built in about 1661 by William, later first earl of Craven (1608–97). Craven's father was a wealthy merchant and a Lord Mayor of London who died when William was only ten. His mother increased the already substantial family wealth by purchasing large estates, including Hampstead Marshall also in Berkshire. She died in 1624 but in 1626 William's trustees bought the manor of Ashdown on the eastern edge of the Lambourn Downs, just south-east of Swindon. The manor included a large well-wooded deer park, created in the fourteenth century by the abbots of Glastonbury. William Craven was knighted in 1626 and created Lord Craven in 1627 when still only nineteen. For the next thirty years he was a soldier on the continent and it was there that he met and greatly assisted the widowed Queen Elizabeth. He was exiled from England during the Civil War, but returned in 1660 and began to rebuild his houses at Hampstead Marshall and Ashdown. He was created earl in 1664 and died, unmarried, in 1697.

Ashdown House is a tall compact building with a cupola and a balustraded viewing platform on the roof. In fact, it is a simple hunting lodge and retreat and it stood within the medieval park and specifically at the intersection of four straight rides in the centre of a large sub-rectangular wood. From its upper storeys and roof the hunts in the adjacent park could be seen. In this view, from the south, one of the rides and the remnants of two others are visible as is the original eastern side of the wood. The tiny formal garden to the left of the house is a modern addition of 1955.

R. Bisgrove, *Gardens of Britain* 3 (London: 1978), pp. 25–6; J. Kip and L. Knyff, *Britannia Illustrata* (London: 1707, reprinted 1984), plate 46; neg. no. CPS 5, 15 October 1982.

Chapter 6

The Landscape Park

Of all the types of parks and gardens that exist in Britain it is the landscape park that is perhaps best known and which figures large in the modern countryside.[1] There are few places in England where there is not a landscape park within a few kilometres and they exist too in considerable numbers in Scotland and Wales. They range in size from the very large, as at Woburn **[77]**, to tiny pocket-handkerchiefs, as at Hackthorn **[34]**. They are associated with great palaces, such as Holkham **[35]**, or modest country houses, such as Cannon Hall **[36]**. Most are regarded as eighteenth-century but, as was noted earlier, parks of various forms had existed since late Saxon times and they continued to be created until the end of the nineteenth century. Further, many later eighteenth-century parks were not new. Some were centuries old and merely updated.

A number of these older parks had formal arrangements of plantations and avenues by at least the early seventeenth century and this geometric style was much reinforced by the fashions of the late seventeenth and early eighteenth centuries. However, from the 1720s, at least among persons of 'taste', there was a growing demand for parks to be more 'natural' – even if 'natural' was relative. Although designers such as William Kent (1685–1748) both advocated and actually laid out lakes with curvilinear sides, and 'serpentine' rivers and walks, the planting of parks remained as formal as ever, as at Heythrop **[28]** and Horton **[37]**, until the 1730s and 1740s.[2]

Yet within thirty years the majority of parks had been changed completely and had become 'natural' landscapes. They were characterised by irregular or curved-edged plantations, belts of trees along the boundaries, small circular clumps, meandering drives and, if possible, a serpentine lake. The intention was to have an apparently natural scene with numerous and continually changing views from both house and drive as exemplified by Weston [38]. Temples, towers, obelisks and statuary, classical, gothick or oriental in design, were placed to add variety or to punctuate or close vistas, as at Boconnoc [39].

To achieve a natural appearance in such parks required detailed survey and planning, civil engineering work, the planting of semi-mature trees, and, most of all, considerable design ability. However, although many parks did indeed have a skilled designer, as at Weston [38], Attingham [40] and Shardeloes [41], many such as Hackthorn [34], Horton [37] and Great Barton [76] did not. The best-known landscape gardener was 'Capability' Brown (1716–83) who created around 170 parks as well as being a talented architect.[3] His works included here are Weston [38], Trentham [42] and Appuldurcombe [75]. Brown's reputation derives as much from his activities as a businessman as from his skill as a designer. He not only created wonderful parks but also ensured that he had important and influential people as clients. However, there were a number of other gifted landscape gardeners of the period whose works, if not better than those of Brown, often came close to or equalled his. Illustrated here are those of William Emes (1730–1823), who carried out minor changes at Holkham [35], Thomas Leggett (fl. 1780–1810), an Irish landscape gardener who created Attingham [40], Nathaniel Richmond (?–1785) who was responsible for part of Shardeloes [41], Richard Woods (1716–93), the designer of Cannon [36] and Adam Mickle (c. 1730–1809) who altered Tredegar [74].[4]

There were also a number of extremely talented amateur landscape designers, some of whose creations matched or outshone anything by professionals. Among these were Thomas Pitt (1737–93) and his uncle William Pitt, Earl of Chatham (1708–78), who together probably created Boconnoc [39]. In addition, there are parks which, although rarely documented, seem to have been made by their owners, perhaps purely for reasons of fashion. Most of these, by their apparent lack of conscious designs, indicate that no great creative genius lay behind them although, in fact, they actually constitute the majority of parks that were both laid out and survive. Hackthorn [34] and Great Barton [76] are examples. Holkham [35] seems to be in a special category. One of the greatest of all landscape parks in Britain, possibly its design and certainly its creation seem to have been largely the work of John Sandys (fl. 1770–1808), Thomas Coke's head gardener. Yet it may be that Holkham is less unusual in this respect than is generally supposed and that only a lack of documentation or proper study prevents others being recognised elsewhere.

As with most parks and gardens, these Brownian-type parks did not remain fashionable for long. Among people of 'taste', as well as in the view of professional designers who were always eager to persuade clients to make changes, the natural landscapes designed by Brown and his contemporaries were soon superseded. Even in his own lifetime Brown was criticised for his bland landscapes. New ideas continued to appear and were quickly accepted by intellectuals and wealthy landowners. These ideas included the recreation of elaborate gardens around the great houses but, in fact, this practice had never actually been abandoned. Both Brown and his contemporaries often laid out formal and informal pleasure grounds near the great house. These usually comprised walks, pools, buildings, woodlands and flower-beds such as still exist at Weston [38] and Cannon [36]. Subsequently, more extensive and elaborate gardens near the houses were advocated, together with parks that had more varied and more 'picturesque' planting. This latter concept was in part encouraged by the increasing numbers of exotic tree species then arriving from both the old and the new worlds and which plantsmen were keen to grow and nurserymen keen to sell.[5]

The most successful of the new landscape designers was Humphrey Repton (1752–1818).[6] His works illustrated here are Attingham [40], Caerhays [60], Shardeloes [41] and Woburn [77]. At first sight these landscapes seem little different from earlier ones. The perimeter belts and plantations were still there, as were the lakes. Even the pleasure grounds near the house, although somewhat more formal and colourful in tune with the new ideas, continued, but Repton created carefully contrived views through the boundary belts to the countryside beyond. And this countryside was sometimes further enhanced by the planting of other clumps of trees. The result was a 'borrowing' of the outer landscape or the drawing of this countryside into the park. This often gave the impression of an even larger estate, as at Attingham [40]. This was particularly important for lesser landowners with small parks who needed their social status enhanced in whatever way possible, for Repton and his supporters were not concerned merely with aesthetics. Status and image were vital for his clients, and thus also for him. Carefully stage-managed park entrances and views of churches were also part of the overall enhancement of parks. In the end, inevitably, Repton himself was overtaken by other changes in fashion, in particular by the further development of the picturesque style. This required yet more variety and detail in the form of crags, waterfalls, quarries, grottoes and bridges, together with lusher and even more exotic planting in the place of the traditional native hardwoods. The later stages of the development of Studley Royal [31] are an early example of the picturesque. This was connected with complex social, political and aesthetic developments that were also widely disseminated.[7]

In addition to the philosophical and political issues which underlay the picturesque movement, and quite separate from, or at least only marginally connected to them, was a continuous process of creating simple landscape parks in a very general Brownian style that continued through to the end of the nineteenth century. The intelligentsia might argue over the finer points of taste or design but the smaller landed gentleman, the *nouveau riche* industrialist or banker, and even some of the major landowners, continued to demand and have open, semi-natural landscape parks. These, separated from the outside world by boundary belts, with gate lodges to control ingress and all giving the requisite appearance of wealth and a solid social position, could be easily produced and cheaply maintained. Few of these nineteenth-century parks are particularly well designed and most, by the standards of the eighteenth century, are poorly laid out. Horton **[37]** is one such instance. However, they are significant both in their numbers and for the way that they indicate the aspirations of the society that created them. Here Croxton **[80]**, laid out in the 1820s by the grandson of a successful Liverpool merchant, is a good example. In this case the existing hedgerow trees were incorporated to help to produce an instant park and not even the hedgerow banks themselves were removed.

Most existing parks were also modified in the nineteenth century, particularly by the planting of exotic trees which were arriving from all over the world. The planting was sometimes designed and sometimes not, although the results were usually impressive.[8] The modifications ranged from the extensively lavish, as at Penrhyn **[50]** in the mid-nineteenth century, to the relatively minor at Stretham Hall of 1878 which formed the nucleus of the later campus of the University of Exeter **[71]**. Other parks were greatly extended, not necessarily by professional designers, nor even particularly well, but often as part of a continuing desire for visible signs of ever-increasing status. The northern part of the park at Woburn **[77]** is one such massive expansion.

Most of this chapter has inevitably dwelt on the importance of taste, authority and status that lay behind the desire for landscape parks. While it is certainly true that these parks did indeed reflect, however dimly, the beliefs and philosophies of contemporary society, it is also true that occasionally there are records of the simpler pleasures that they provided. The diary of John Spencer, the Sheffield ironmaster and owner of Cannon **[36]**, makes this clear. Although he employed Richard Woods to create his park and lake, he was deeply and personally involved in all stages of its development. And finally, on its completion and on his return from the hectic London season in July 1765, he happily wrote 'At Home all Day sailing and fishing'.[9]

34 Hackthorn Park, Lincolnshire

Hackthorn is more typical of eighteenth- and nineteenth-century parks in Britain than any other in this book. It is small, only 40 hectares in extent, of no artistic merit and had no designer. Yet parks such as this dot the English landscape and give it much of its character. It lies 12 kilometres north of Lincoln on an almost level site, with no spectacular views. It is seen here from the east, with the Hall and the parish church close together, centre right. Most of the village of Hackthorn lies off and below the picture, but an old hollow-way with traces of former houses along it can be seen immediately south, left, of the narrow lake, indicating that the village once extended into this area. This part of Hackthorn was abandoned in the fourteenth century. Elsewhere in the park the ridge-and-furrow of the medieval fields is visible.

In the late seventeenth century the manor of Hackthorn was purchased by the Cracroft family, originally from Whisby, south-east of Lincoln. The Cracrofts moved here before 1677 and may have rebuilt the manor house and possibly created a small park. The present park, however, dates from the late eighteenth century. The enclosure of the medieval fields of Hackthorn by an Act of Parliament in 1788 and their replacement by rectangular hedged fields gave the Cracrofts the opportunity to create a park and thus a better setting for their house. The new allotments of land were arranged so that the family secured a large compact block of land around the house and they immediately laid out the park in 1778–9. Its rectangular shape followed the new fields, it was given very thin boundary belts and trees were planted randomly across it. The modest lake, if such it can be termed, was created by damming the tiny Hackthorn Brook which rises only a short distance to the west. In 1792 John Cracroft rebuilt the house in a restrained classical style and it was further enlarged in 1840 by the Amcott family who succeeded the Cracrofts.

P. Everson, C. Taylor and C. Dunn, *Change and Continuity* (London: 1991), pp. 106–7; N. Pevsner, *Lincolnshire* (2nd edn, London: 1989), p. 358; neg. no. BZU 46, 13 July 1976.

35. Holkham Hall, Norfolk

On the north Norfolk coast, and here viewed from the south-east, is one of the greatest eighteenth-century parks of Britain. The Holkham estate was acquired in 1612 by Sir Edward Coke (1552–1634), Lord Chief Justice and MP. The Coke family lived in the old manor house at the side of the ancient village green, on the site of the present Palladian mansion. In 1707 the estate, together with a fortune, was inherited by Thomas Coke (1697–1759) who was created Earl of Leicester in 1744. After he returned from the Grand Tour in 1718 Coke began to plan a new house and park, with the aid of Lord Burlington (1695–1753), William Kent (1685–1748) and Matthew Brettingham (1699–1769). The house was started in 1735 but not completed until after Coke's death. The park was finished before 1759, perhaps to designs by Kent, although Coke was probably involved. Its creation was facilitated by the removal of the old village and its fields. It was entirely formal with radiating avenues extending in all directions from rides cut in a grove to the south of the house and its formal garden. This park covered only the north-central part of the present one and all its layout has been obliterated. Its 3.5-kilometre long southern approach drive and avenue survive and are incorporated within the later park, in the foreground.

The present park belongs to the later eighteenth century and was begun in the lifetime of Coke's nephew, Robert Wenham (1717–76), who took his uncle's name when he succeeded. In 1762 Robert Coke employed Capability Brown but the modifications that he made were limited. When his son, Thomas William Coke (1754–1842), the great agricultural improver 'Coke of Norfolk', succeeded in turn, he began the development of the park to its present size (1200 hectares) and appearance. Some work was done in the early 1780s by William Emes (1730–1803) who enlarged Kent's lake, just visible in the photograph above and to the left of the hall, but the enormous task of closing roads, removing hedges, erecting new buildings and, above all, planting an estimated two million trees in clumps, belts and as individual specimens, was largely the work of John Sandys (fl. 1770–1808), Coke's head gardener. Sandys came to Holkham, perhaps with Emes, about 1780 and left in 1805, and it is he who can be credited with both the design and the layout of Holkham. Even Repton, who was employed here in 1789, contributed very little.

A. Davison, 'Six deserted villages in Norfolk', *East Anglian Archaeology* 44 (1988), pp. 71–82; T. Williamson, *The Archaeology of the Landscape Park*, British Archaeological Reports (1998); neg. no. FQ 78, 15 June 1951.

36 Cannon Hall, Yorkshire, West Riding

Cannon Hall lies seven kilometres north-west of Barnsley. The house is of late seventeenth-century date and originally had a formal garden in front of it. It was owned by the Spencers who, from modest beginnings, rose to become one of the major iron-making families of Sheffield. In 1759 John Spencer, a leading ironmaster, commissioned the then rising landscape designer Richard Woods (1716–93) to lay out a new park. The work began in 1760 and continued until 1765 when Spencer began to enlarge the house.

Spencer's diary, which survives, is of considerable interest and shows how Woods worked in close collaboration with his client and did not merely impose his ideas on him. The diary also details Woods's methods which included marking out proposed tree clumps and lakes with stakes, thinning standing timber and moving trees. The final landscape, although Brownian in concept, was not in the Brown tradition in that Spencer had a garden laid out around the Hall consisting of winding paths, a pond, flower-beds, rose borders and lawns, all within a ha-ha.

The view here is from the north-east and shows only the centre of the park. Much of its southern part is now arable and the Hall lies just off the photograph, to the north-west, right, on rising ground. The park, which covers about 150 hectares, is modest in its arrangement with individual trees or clumps scattered across the landscape. Belts of trees lie along the edges and on the south-west screen off a public road. A small brook has been dammed with a series of cascades to create three sinuous lakes which are crossed by two drives on bridges. These bridges were described as Palladian but are in fact simple arched and balustraded structures. One is visible in the left foreground. In 1765, when the work was complete, Spencer ordered two boats for 'my river' and to his great delight was soon sailing and fishing on the new lakes.

B. E. Coates, 'The work of Richard Woods', *Transactions of the Hunter Archaeological Society* 8 (1963), pp. 298–306; F. Cowell, 'Richard Woods', *Garden History* 14.2 (1986), pp. 85–119; A. Raistrick and E. Allen, 'The South Yorkshire iron masters', *Economic History Review* 9.2 (1939), p. 154; G. Sheeran, *Landscape Gardens in South Yorkshire* (Wakefield: 1990), pp. 50–4, 193–4; neg. no. CQB 72, 8 June 1984.

37 Horton, Northamptonshire

Horton Park lies 10 kilometres south-east of Northampton. Here, seen from the east, is one section of an undistinguished park, the designer of which is unknown. The earliest visible feature is the corrugated ridge-and-furrow which marks the medieval open fields of Horton village. The latter had two separate parts. One, together with the site of the Hall, still exists to the west, beyond the top of the photograph. The other part, disposed around a small green, lay south-east, half-left, of the central circular feature. Its site is marked by slight undulations and is cut across by a later ditch.

In the early seventeenth century Horton was bought by Henry Montagu (c. 1563–1642), later first earl of Manchester. Henry's grandson George (c. 1685–1739) inherited the title of Earl of Halifax from his uncle in 1715 and at once began to rebuild the Hall and to create a grand geometric park. He removed that part of the village which lay in the area of the intended park and planted north and south of the house long avenues comprised of multiple rows of trees similar to those at Heythrop [28]. A third avenue to the east was never planted but its surveyed lines are marked on the ground by the parallel ditches extending across the ridge-and-furrow in the foreground. The circular feature at the west end of these ditches was intended as a pond. The park was far from complete in 1728 when a map was made of it and it was in fact never finished.

When George Montagu (1716–71), the third earl, Lord Lieutenant of Ireland and Lord of the Admiralty, succeeded in 1739, he completed the house and began a new informal landscaped park. This involved the removal of the avenues, the creation of the rather meagre lake, top left, and the planting of the boundary belt in the bottom of the picture. In 1781 Horton was bought by Sir Robert Gunning (1731–1816) and remained in his family until the late nineteenth century. Among the changes made by the Gunnings, probably in the early nineteenth century, was the building of the Triumphal Arch in the centre foreground which was originally flanked by arcs of trees set within curved ditches. The ditches and the tree holes are visible on either side of the arch. Other features visible include a straight ditch, upper right, which supplied water to the Hall gardens, and a number of mounds and ditches on the far side of the lake. The date and purpose of these are unknown but local tradition suggests that they were where the third earl exercised his private militia.

RCHME, *Northamptonshire* II (London: 1979), pp. 65–71; J. Heward and R. Taylor, *Country Houses of Northamptonshire* (London: 1996), pp. 238–40; neg. no. BKZ 46, 23 November 1972.

38 Weston Park, Staffordshire

Weston Park lies eight kilometres north-east of Shifnal on the south side of the A5, Watling Street. In this view, from the south-west, only the eastern third of the large 300-hectare park is visible. The house and the rest of the park lie to the left. The late-seventeenth-century house and a smaller park came to the earls of Bradford by marriage. Sir Henry Bridgeman (1725–1800) inherited Weston from his maternal uncle, the fourth earl, in 1762 and was himself created the first Lord Bradford in 1794. In 1765 Bridgeman engaged Capability Brown to enlarge and landscape the park and to create pleasure grounds near the house, all of which is said to have cost some £12,000. The work included the creation of two lakes and James Paine (c. 1716–89) designed a Roman bridge and a Temple of Diana in 1770.

In the photograph, this part of the park appears extensively planted as is its western half. The central section, to the south of the house, is more open, with broad views. Here there are individual trees, irregular clumps, curving belts and plantations which, from both the house and the sinuous drives, provide glimpses of long and short views in a constantly changing pattern. The open space in the top centre is a smaller open glade through which the main approach drive from the east passes. Although inevitably somewhat altered over the succeeding two centuries, Weston Park retains much of its original layout.

C. Hussey, 'Weston Park', *Country Life* 98 (1945), pp. 910–13; R. Sidwell, *West Midland Gardens* (Gloucester: 1981), pp. 155–7; D. Stroud, *Capability Brown* (London: 1975), pp. 148–9; VCH, Staffordshire IV (London: 1958), pp. 170–3; neg. no. FA 26, 10 July 1950.

39 Boconnoc, Cornwall

Boconnoc lies four kilometres north-east of Lostwithiel on land cut by two steep-sided valleys of the River Lerryn. It is a surprisingly ill-documented park of considerable beauty. An existing house and park was purchased in 1717 by Thomas Pitt (?–1761), former Governor of Madras, who extended the house with the considerable proceeds of the sale of the Pitt Diamond brought back from India. His grandson, also Thomas Pitt (1737–93), first Lord Camelford and nephew of William Pitt (1708–78), Prime Minister and later Earl of Chatham, inherited Boconnoc and further extended the house. In the 1760s he improved the park by the creation of extensive plantations, by the laying out of many kilometres of carriage drives and by the erection of a number of classical features including an obelisk. His uncle may well have advised him on the landscaping for William Pitt was an amateur landscape designer of ability, was involved with a number of gardens and was a friend of Capability Brown. Thomas Pitt was succeeded by his son, another Thomas (1775–1804), who, as an early exponent of the picturesque, probably further enhanced the appearance of the park and its surrounding woodland.

Lord Camelford was killed in a duel and Boconnoc passed to his sister Anne, wife of Lord Grenville (1759–1834), sometime Prime Minister. Grenville made further additions to the estate, including a pinetum. On Lady Grenville's death in 1865 Boconnoc passed to the Fortescue family in whose possession it has remained.

In the view here, from the south, the central open parkland with scattered clumps and individual trees is visible. So too are the extensive beech woods which fill the valleys and extend on to the interfluves as belts or plantations and through which the carriage drives still pass. On the extreme left is the eastern end of a separate deer park, also bounded by plantations, which extends a further two kilometres north-west. In the woodland to the north-east, upper right, the obelisk can be seen.

G. S. Gilbert, *Historical Survey of Cornwall* (Plymouth: 1820), pp. 908–12; S. Pring (ed.), *Glorious Gardens of Cornwall* (Oxford: 1996), p. 54; neg. no. BON 51, 26 July 1973.

40 Attingham Park, Shropshire

Attingham lies five kilometres south-east of Shrewsbury, the view here being from the north-east across the valley of the River Tern. The house is visible in the middle distance. An early-eighteenth-century house, called Tern Hall, was probably built by Richard Hill who bought the property in 1701. Hill was the grandson of John Harwood, a draper, and Mayor of Shrewsbury in 1680. Hill's father, Thomas Harwood, married into the Hill family of Hawkestone, Shropshire, took the family name and began its rise into the aristocracy. Noel Hill (1745–89), MP for Shrewsbury and Shropshire 1768–84, and son of Richard Hill, was created Lord Berwick in 1784. He began to prepare for his elevation to the peerage before he achieved it, for between 1769 and 1772 he employed a Thomas Leggett to create a landscape park. Leggett may have been Thomas Leggett (fl. 1780–1810), the leading eighteenth-century Irish landscape designer. Large sums were spent on planting trees and shrubs and changing the valley of the Tern. The river was also deepened. It was Lord Berwick who, in 1783–5, rebuilt the old hall, presumably as an establishment appropriate to his new title. His son, Thomas Noel-Hill (1770–1832), further enhanced the setting of the house. In 1797 he called in Humphrey Repton (1752–1818) whose proposals were partly implemented.

Repton worked on the existing parkland, modifying it rather than changing it radically. He widened the River Tern into a large lake although this was subsequently drained and only its marshy bed and a straight channel are now visible below the house, in the middle distance. He introduced trees in clumps and plantations either from new or by uniting smaller groups of trees to create the effect of the house and its immediate surroundings being within a limitless park. Repton also opened gaps in Leggett's shelter belts to give views to the landscape beyond. He further proposed extending the park westward to give greater privacy, and this was eventually achieved, although it involved the removal of the hamlet of Berwick Maviston. Attingham's landscape suffered from neglect in later times, but since its acquisition by the National Trust in 1953 the Repton layout has been restored.

P. Stamper, *Historic Parks and Gardens of Shropshire* (Shrewsbury: 1996), pp. 57–65; neg. no. FA 36, 10 July 1950.

41 Shardeloes, Buckinghamshire

Shardeloes lies three kilometres west of Amersham in the valley of the Misbourne. The medieval house here was bought in 1593 by William Tothill, a clerk in Chancery, who added a new front to it. In the early seventeenth century it passed by marriage to the Drake family of Esher, Surrey. The Drakes settled at Shardeloes and became a county family, controlling the two Amersham parliamentary seats and often being elected as MPs. Shardeloes then had a very formal garden with a canal, pools and an equally formal park. In 1728 Mountagu Garrard Drake (?–1728) proposed rebuilding the house but nothing came of it until his son, William Drake (1723–96), also MP for Amersham, returned from a Grand Tour in 1758. He originally employed Stiff Leadbetter to design the new house but it was altered and completed by Robert Adam in 1766. It was presumably at this time that the formal gardens and park were swept away and a more natural landscape established. Some of this was done by Nathaniel Richmond (?–1785), landscape designer and follower of Capability Brown. In the 1760s he created the large triangular lake formed by damming the Misbourne and visible on the left of the photograph which is taken from the west. However, what else he did at Shardeloes is unclear.

In 1793 Drake, or more likely his son, also William (1747–95), called in Humphrey Repton (1752–1818) to redesign the park, although again what he did is not exactly clear. Certainly the large plantation at the east, far, end of the lake is his work, as is much of the tree planting in the north of the park, off the photograph to the left. The large plantations south and west of the house, as well as the copses along the valley side, may also be Repton's. Although there have been changes to the area immediately around the house in both the nineteenth and the twentieth centuries, Shardeloes remains a particularly well preserved late-eighteenth-century landscape.

G. Carter, P. Goode and K. Laurie, *Humphrey Repton* (Norwich: 1982), p. 140; J. Harris, 'Shardeloes', *Connoisseur* 148 (1961), pp. 268–75; neg. no. CMV 4, 24 July 1980.

Chapter 7

Nineteenth-century Gardens

The history of gardens in the nineteenth century is more complicated than in earlier times.[1] As always, the gardens were reflections of the society that created them and nineteenth-century society, or at least that small part of it that could afford to create gardens, was complex, changed rapidly and was full of contradictions. It was, on the whole, a society that was assured, self-confident and even brash and it was underpinned by great wealth from land, commerce, and especially, industry. This wealth came to both the ancient aristocracy and the *nouveau riche* bankers and industrialists. At the same time, political power moved from a basically conservative landowning élite to a somewhat more liberal, democratic and largely urban populace, all of which affected gardens in various ways.

More directly, gardens were affected by new technology, including greenhouses which aided propagation and overwintering, lawnmowers which made the management of grass easier, new tools, drainage methods and building materials. A major influence was the arrival of new plants from abroad, which poured in as a result of a steadily increasing number of expeditions to the New World and the Far East, and improved methods of transport. A result of all this was that gardening, in all its forms, became more professional. Head gardeners, assisted by large numbers of staff, were permanently employed in most of the bigger gardens. They had to cope with all the new plants, new techniques and new demands. Far too little is known of these people although they were often extraordinarily influential. Only occasionally, as at Trentham [42], Levens [27] and Drummond [43], do they surface.

Another development was the enormous increase in garden experts, whether professional designers or writers, or amateur plantsmen or publicists. Such people had, of course, existed since at least the medieval period, but now there were many more of them. The advent of cheap printing methods contributed to a flood of books and, in particular, to the establishment of journals such as the *Garden Magazine* (1826), the *Gardeners' Chronicle* (1841) and the *Ladies' Magazine of Gardening* (1841). The founding in 1804 of the Horticultural Society of London, later the Royal Horticultural Society, also did much to stimulate interest in and knowledge of gardens and garden design.

All these developments produced more gardens than ever before, but also more diverse gardens in terms both of design and of planting. Fashions came and went with great rapidity and arguments over styles, methods of planting and types of plants dominated the available literature. It is difficult for the garden historian to chart a coherent passage through all the changes and for the landscape historian it is even worse. Although there is much information on nineteenth-century gardens, both written and illustrative, it remains surprisingly difficult to point to any reasonably complete gardens still visible today. Few, if any, remain as they were and, even if they did, the evidence of old photographs suggests that they would not be popular.[2] By the today's standards they usually appear to be extraordinarily cluttered.

For much of the nineteenth century, perhaps in part as a reaction to the 'natural' landscapes of the later eighteenth century, most garden designers seem to have followed Repton's precept that 'Gardens are works of art rather than Nature'.[3] That is, many nineteenth-century gardens were formal in arrangement and were often claimed to be based on those of earlier periods, both in Britain and abroad. However, the perception of what past gardens had actually looked like varied almost from decade to decade and, combined with the desire to incorporate new plants and materials into gardens and garden designs, meant that historical accuracy was only rarely achieved. The only 'style' that remained reasonably popular throughout the century was what could be described roughly as, and sometimes was, Italianate, but this style coexisted with many other shorter-lived fashions for Renaissance, Jacobean, French, Dutch, Chinese, Japanese and even Swiss gardens. It is impossible to illustrate here even a fraction of the range of forms of nineteenth-century gardens and indeed many are poor subjects for aerial photography.

One of the earliest nineteenth-century examples of an attempted return to the past can be seen at Levens [27], where Alexander Forbes (fl. 1810–60) was head gardener. The long-neglected later seventeenth-century topiary there was recut in about 1815 by Forbes who also added much new box edging. The topiary was almost certainly recut incorrectly and some entirely new figures were introduced. Levens is thus as much a recreation of the early nineteenth century as a seventeenth-century original.

An early example of a virtually new 'historical' garden, albeit on an old site, is Drummond **[43]**, created in the 1820s and 1830s. It was meant to be in a seventeenth-century Scottish style, although with rhododendrons and heathers in the knots it was hardly that, and later writers described it variously as Italian, Dutch and French.[4] A later Scottish garden, of the 1850s, is Drumlanrig **[44]**. There a late-seventeenth-century structure was reused but again heather was planted, here within complex parterres based on late-seventeenth-century French designs. This use of elaborate French-style parterres, sometimes comprising box and coloured gravels alone, became popular in the 1840s and 1850s and most notably was used by the designer W. A. Nesfield (1793–1881).[5] Although some of Nesfield's early layouts were relatively simple, as at Somerleyton **[45]** in the late 1840s, some of his subsequent ones were extremely elaborate and often divorced from historical reality. One such is Witley Court **[46]**.

In contrast to differences in planting schemes there was, in the mid-nineteenth century, a growing uniformity in the underlying structures of gardens. These came to be more and more Italianate in the widest sense, with terraces, balustrading, elaborate stairways and wide vistas. This was due in part to the popularity of Italianate architecture for country houses, deriving in particular from Prince Albert's work at Osborne **[47]** but it also had a much longer history that stretched back to the sixteenth century and which was given added impetus by the very influential Italianate house and garden at Deepdene, Surrey, of 1818–23 by Thomas Hope.[6] Trentham **[42]** of the 1830s and Osborne **[47]** in the 1850s are two different interpretations of the Italianate style, and Thoresby **[48]**, nearly twenty years later, is another.

As the century advanced, the actual structures of gardens tended to be swamped by elaborate planting. As early as the 1830s and 1840s there was an increasing enthusiasm for brightly coloured ornamental bedding, often using newly introduced exotic plants, as at Wimpole **[49]** and Penrhyn **[50]**. By the 1850s flower-beds had become the centres of attraction, growing increasingly more exotic, colourful and complex in design. These, and the even more elaborate carpet bedding, remained popular until the end of the century and beyond, alongside newer and very different fashions of planting. Even in the 1860s there was a reaction against highly coloured bedding and an increasing use of 'native' plants in more restrained colours. This again was often justified on historical grounds, the plants used being those described by earlier writers such as Milton, Shakespeare and Spenser, and thus assumed to have been associated with historic gardens. From this emerged a style of gardening, and associated architecture, often known as English Vernacular. This style involved the creation of broad herbaceous borders, the exact form and contents of which were debated by William Robinson (1834–1935) and Gertrude Jekyll (1843–1932) among others.[7] Lawns, now that they could be mown easily, grew larger and formed part of long views. This aspect is visible in some of the larger early-twentieth-century gardens such as Mellerstain [56] and Dyffryn [58]. The general relaxation in formality also extended to the creation of shrubberies, ferneries and rock gardens, with winding paths and exotic trees, which, when accompanied by broad lawns, produced the gardens so favoured by the middle-class owners of late-nineteenth-century villas, as at Stretham Hall, Exeter [71].

One other long-lasting and hugely influential development of the nineteenth century was the compartmentalisation of gardens. For the keen plantsman and gardener the nineteenth century represented utopia. New plants were widely available and information about gardens and garden designs from all over the world was easily obtained. The difficulty was to use this information and to put all the new plants into an arrangement that bore some relation to their history, original location and habitat, or to produce a suitably authentic layout. The first attempts produced wonderfully disordered schemes, most famously at Alton Towers, Staffordshire, in the early years of the century.[8] The problem was solved, notably by James Bateman (1811–97) in the 1850s at Biddulph Grange, also in Staffordshire, now restored almost to its original condition.[9] Although by no means new, Bateman's idea was to create separate compartments, cleverly linked to achieve an overall unity, containing a Chinese, an Italian and an American garden, a pinetum, an arboretum and so on. This concept was taken up by other garden designers, in particular by Edward Kemp (1817–91) at gardens such as Leighton [51] in the 1860s. By the 1890s many gardens were being given walled or hedged enclosures as part of their basic structure, perhaps the best example being Athelhampton, Dorset, designed by Inigo Thomas (1866–1950). The concept was further developed in the twentieth century to produce gardens such as Dyffryn [58], Great Dixter [59] and Sissinghurst [53].

The motives behind the creation of this great variety of garden forms in the nineteenth century appear to have been as mixed as ever, although the need for as ostentatious a layout as possible is most obvious. Indeed, this may have contributed more to garden designs in the nineteenth century than some garden historians have admitted. Certainly the size of Nesfield's parterres at Witley [46], together with the house itself, were necessary for the great social gatherings held there. Anything smaller simply would not have done. The same is probably true of Trentham [42]. For those whose new wealth enabled them to rise into the landed classes, a major garden was as essential as a new house. Peto, the railway contractor at Somerleyton [45], and Naylor, a banker at Leighton [51], both had fine gardens made for them by top-rank designers.

There were also, of course, other motives for gardens. The extensive literature on gardening shows the growing interest in every aspect of gardens by the new and numerous middle classes. More specifically, the gardens at Osborne [47], although lavish in detail, are relatively small and perhaps reflect something of the character of Prince Albert. Although no great gardener, he seems to have created at Osborne something closer perhaps to what he desired rather than what might have been thought suitable for the palace it adjoined. And, although only just nineteenth-century in date, Batsford [52] stands for a whole group of gardens and landscapes made by those whose interest in and love of plants outweighed the fashionable exigencies of contemporary society.

42 Trentham Park, Staffordshire

The site of the medieval priory of Trentham was bought by the Leveson family, rich Wolverhampton wool merchants, who built a house there in the 1630s. The Levesons continued to prosper socially and financially so that, by the late eighteenth century, they were marquises of Stafford with extensive Midland estates. The second marquis, George Grenville Leveson-Gower (1758–1833), married Elizabeth, Countess of Sutherland (1765–1839), who brought to her husband vast tracts of the Scottish Highlands as well as, later, the dukedom of Sutherland. The duke was an agricultural improver and certainly increased the value of his English lands but he gained lasting notoriety by his eviction of up to 15,000 crofters from his Sutherland estates. Most of the family's wealth, however, came from the coal and iron below their Staffordshire and Shropshire properties. By 1883 they owned more land in Britain than any other aristocratic family and were the fifth richest.

When George Grenville, the second duke and later first earl (1786–1861), inherited Trentham in 1833 he altered the house and created new gardens. The old house had already been remodelled and Capability Brown had laid out a new park which still survives. The duke engaged Sir Charles Barry (1795–1860) to make the house an Italianate palace and to place between it and Brown's great lake an elaborate parterre.

In this view from the north-west, the site of the house, with its surviving curved *porte-cochère* to the right and its orangery to the left, lies in the foreground. Immediately above it is the square Upper Garden with a fountain in the centre. At its north end is a balustrade with steps leading down into the Lower Garden. Originally there were four pavilions at the ends of the balustrade, two on each side, although only part of one survives. The Lower Garden comprises six panels, planted in complex patterns, each one formerly with a central fountain. Lining the main axial path and the side paths are Portuguese cherry laurels planted in pseudo-tubs and trimmed to look like orange trees. At its far end, on a projection into the lake, is a statue of Perseus. On either side of this garden are the remains of ornamental pleasure grounds.

Although the architectural framework of the garden was by Barry and was created between 1833 and 1841, the planting was by the duke's head gardener, George Fleming (1809–76), who arrived in 1841. The Sutherlands remained at Trentham until the end of the century and in 1910–11 most of the house was pulled down.

B. Elliot, *Victorian Gardens* (London: 1986), pp. 75–7, 90–4; D. M. Palliser, *The Staffordshire Landscape* (London: 1976), pp. 129–33, 228–9; R. Sidwell, *West Midland Gardens* (London: 1981), pp. 177–9; neg. no. AAZ 92, 14 June 1960.

43 Drummond Castle, Perthshire

Drummond lies four kilometres south-west of Crieff. Although the castle dates from 1491 the gardens are of the nineteenth century. They were among the first to be recreated in the then perceived ideal of seventeenth-century Scottish gardens.

The Drummond family, who held the castle in later medieval times and who already had a garden there in 1508, were strong supporters of the Scottish Crown. James Drummond (c. 1580–1675) was created earl of Perth in 1604 because of this support. Soon afterwards he built a house alongside the castle and later laid out new gardens on the site of the present ones. These consisted of four square knots. James, the second earl (1648–1716), followed James II into exile and both the next two earls also supported the Jacobite cause. As a result their estates were neglected and then forfeited in 1745. They were not recovered until 1785 when the family abandoned its ancient allegiance. However, it was not until 1818 that Lewis Kennedy (1789–c. 1840) followed later by his son George (c. 1820–c. 1860), successive factors to the Drummond estates and influential garden designers, began to lay out the gardens as they now exist. The work was not completed until the 1840s.

In this view from the east, the medieval tower and its successor stand on the crest of a ridge, to the right, with the gardens cut back into the hillside. Below the buildings is a simple balustraded terrace from which modern stairs lead down across a sloping terrace to the main garden. This is an elongated rectangle divided into compartments by gravel paths and overlain by a somewhat stretched St Andrew's cross of grass. In the centre is a reset sundial of 1630. The present planting, although apparently complicated, is actually much simplified from its original, or even its early-twentieth-century arrangements. In its heyday the compartments were filled with a confusion of patterned beds containing shrubs, rhododendrons, heathers and herbaceous plants which actually hid most of the structure, except for the paths which were lined with coloured gravel.

'Drummond Castle' *Country Life* 12 (1902), pp. 112–22; B. Elliott, *Victorian Gardens* (London: 1986), pp. 62, 67; A. Hellyer, 'A great Scottish formal garden', *Country Life* 152 (1972), pp. 338–40; neg. no. CKO 46, 1 August 1978.

44 Drumlanrig Castle, Dumfriesshire

Drumlanrig lies 27 kilometres north-west of Dumfries. The castle was given to William Douglas, natural son of James, second earl of Douglas, by his father in 1383. It was destroyed in the Civil War and rebuilt in 1679–90 by William Douglas (1637–95), third earl and later Duke of Queensberry. Like his forbears, Douglas was originally a supporter of the Stuarts but after changing his allegiance in 1688 he prospered. His new house reflected the changing circumstances of Scotland, half castle and half country house, just as Douglas was partly Scottish chieftain and partly Whig grandee.

The surrounding garden and park were begun by Douglas and completed by his son James, the second duke (1662–1711). Little remains of the gardens although the original arrangement of nine squares with the house occupying the central one, the forecourt another, stables a third and the rest laid out in parterres, survives in outline and is visible in this view from the south. At first the house and parterres were bounded by walls with gazebos at the corners of the house platform, and certainly by 1750 the lawn in the foreground was also occupied by parterres set within a curved boundary. The park beyond was also formal with radiating avenues and sinuous paths cut through groves.

This park was destroyed, the gardens ruined and the house abandoned by William, the profligate fourth duke (1725–1810), after he succeeded in 1788. On his death Drumlanrig passed to the dukes of Buccleuch. Charles Scott, the third duke (1772–1819), restored the house, and Walter, the fourth duke (1806–89), probably under the direction of his head gardener and writer Charles M'Intosh (1794–1864), recreated the elaborate parterres and constructed the slopes and ramps that still survive. He also laid out new parterres in the foreground. The parterres were notable for the use of heather rather than bedding plants. These arrangements lasted well into the twentieth century. Both the surviving much simplified parterres, as well as the earthworks and cropmarks of the others, far from being seventeenth-century in date are of the mid-nineteenth century and indicate the contemporary interest in seventeenth-century Scottish garden design.

'Drumlanrig Castle', *Country Life* 12 (1902), pp. 240–8; J. Gifford, *Dumfries and Galloway* (London: 1996), pp. 222–30; M. Girouard, 'Drumlanrig Castle', *Country Life* 128 (1960), pp. 378–81, 434–7, 488–91; H. I. Trigg, *Formal Gardens in England and Scotland* (London: 1902), fig. 42; neg. no. UG 35, 8 April 1957.

45 Somerleyton Hall, Suffolk

Somerleyton, which lies seven kilometres north-west of Lowestoft, was bought in 1843 by Sir Morton Peto (1809–89), an archetypal Victorian self-made man. He was born in relative poverty, apprenticed to his uncle as a bricklayer at fourteen and acquired half the business when his uncle died in 1830. He then enjoyed a meteoric rise as a railway contractor, developer and speculator. Within ten years he had become one of the greatest figures in the construction industry. He was responsible for many major buildings and for railways in Britain, Africa, America and Europe. As a speculator he helped to promote and develop a number of resorts including Lowestoft. He was also MP for Norwich in 1847, a guarantor of the Great Exhibition and was made a baronet in 1855. However, Peto's financial base, always precarious, did not outlast the railway boom. He went bankrupt in 1866 and sold Somerleyton to Sir Francis Crossley, the son of another self-made man who had risen from Halifax weaver to head of a great carpet-making empire.

The house at Somerleyton, seen here from the north-east, was completed in 1851. The surrounding gardens were laid out between 1844 and 1862, certainly in part and possibly entirely, by W. A. Nesfield (1793–1881). What certainly can be attributed to Nesfield is the formal garden on the west front of the house, centre top, set on a broad balustraded terrace overlooking an older park. It originally had intricate parterres, typical of Nesfield, and although it has been replanted in a much simplified arrangement since the photograph was taken, most of the original layout survives as slight earthworks. It is a good example of the way that Nesfield took the then popular seventeenth-century type of parterres and developed them into much more complex arrangements which had no historical basis.

On the north side of the house, in the angle between it and the stables, are the foundations of the Winter Garden, a huge glass-domed structure, pulled down in 1914. To its north-west, right, is the North Lawn, now an informal area containing many fine specimen trees which linked the formal parterre to the park and woodland beyond. Originally it was much more geometric, yet another Nesfield feature. In the foreground is a maze of yew hedges, with a Chinese pagoda on the central mound, all by Nesfield. To its west, right, the former kitchen gardens have a fine group of nineteenth-century glasshouses with ridge-and-furrow roofs.

G. Biddle, *The Railway Surveyors* (London: 1990), pp. 131, 134; B. Elliott, *Victorian Gardens* (London: 1986), p. 72; *Illustrated Guide to Somerleyton* (c. 1992); J. Simmons, *The Railway in Town and Country 1830–1914* (Newton Abbot: 1986), passim; neg. no. BVU 10, 12 August 1975.

46 Witley Court, Worcestershire

The old house and park at Witley were purchased in 1837 by the Dudley family who, from the 1740s, had grown rich on the profits of Midland coal-mines, ironworks and quarrying. In 1846 the Dudley heir, William, Lord Ward (1818–85), began the transformation of Witley into a palace. The house was trebled in size and, to create a suitable setting, W. A. Nesfield was engaged in the 1850s and 1860s to make the gardens. These were intended specifically to reflect the wealth and power of one of the richest men in England. By the time he died, Ward, by then the first earl of Dudley, had realised this ambition.

His son, the second earl (1867–1932), lived in even greater opulence at Witley which became one of the foremost centres of late Victorian and Edwardian society. Edward Prince of Wales, later Edward VII, a close friend of the earl, visited on a number of occasions and the extravagance of the parties there became legendary. The earl's much reduced income after the First World War, as well as the accidental death of his wife in 1920, led to the sale of Witley to Sir Henry Smith (1872–1943), a self-made Kidderminster carpet manufacturer. The house was badly damaged by fire in 1937 and both it and the garden were abandoned. They are now in the care of English Heritage.

The formal gardens, seen here from the south-west, which Nesfield called his 'Monster Work', covered about 3.5 hectares. On the south, steps led down from a terrace, along an axial path, to a vast stone basin with a fountain with statues of Perseus and Andromeda. On each side of the path the remains of large circular parterres, once containing box edging and coloured gravel and surrounded by rectangular beds, can still be seen. Below the fountain, on rising ground, is a semicircular area backed by shrubs and flanked by two pavilions with domed roofs. The whole is enclosed by a balustrade with steps to the parkland beyond. To the east, right, of the house are the remains of a rectangular parterre set below side-walks and terminating in another, smaller, Triton fountain. On each side were originally formal arrangements of clipped shrubs, some of which still survive. This area is also enclosed by a balustrade. To the west, half-left, of the house the huge roofless orangery of 1860 is clear.

S. Evans, *Nesfield's Monster Work* (Worcester: 1994); B. Pardoe, *Witley Court* (Worcester: 1986); neg. no. CJ 81, 18 June 1949.

47 Osborne House, Isle of Wight

Osborne House, a plain eighteenth-century building, was bought in 1845 by Queen Victoria and Prince Albert to provide, in the Queen's own words, 'a place of one's own, quiet and retired', but, unlike Balmoral, relatively close to London. Although the Queen thought the house 'complete and snug', it soon proved to be far too small and even before the purchase was completed the contractor and developer Thomas Cubitt (1788–1855) was engaged to build a new house. Completed in 1851 and perhaps designed by Prince Albert himself, it was conceived as a grand villa in the Italian style. It became the Queen's favourite house and she stayed there regularly until her death in 1901.

The formal gardens, north-east of the house, seen here from the west, were also probably designed by Prince Albert, with the aid of Ludwig Grüner (1801–82), a German sculptor whom the Prince employed as his adviser on the decoration of the house. They were completed in 1853 and consist of an L-shaped balustraded upper terrace with a formal layout and two fountains above a lower terrace bounded by massive masonry walls. The two are linked by a complex double stairway. Like the house they are Italianate and are very close to contemporary gardens in Northern Italy such as those at the Villas Torrigani and Garzoni with which Albert would have been familiar. Both terraces had paths laid originally with a metallic lava of different colours, traces of which have been found in recent excavations. Between these paths were displays of massed bedding. These have been much simplified. Below the gardens the original eighteenth-century parkland, also extensively planted under the direction of Prince Albert, falls away across a broad valley to the sea one kilometre to the north-east, left.

C. Aslet, 'Osborne House', *Country Life* 178 (1985), pp. 1764-7; M. Girouard, *The Victorian Country House* (Yale: 1979), pp. 147-53; M. Turner, *Osborne House*, English Heritage (London: 1989); neg. no. CM 9, 20 June 1949.

48 Thoresby Hall, Nottinghamshire

Thoresby lies 10 kilometres south-east of Worksop. House, gardens and park have a long and complex history as a result of their ownership from the early seventeenth century by the Pierrepoint family, later earls and then dukes of Kingston upon Hull. In 1788 Thoresby passed by marriage to Charles Meadows (1737–1816) who took the name Pierrepoint and was created Earl Manvers in 1806. Meadows's grandson, Sidney William Pierrepoint (1825–1900), the third earl, unlike his predecessors, kept out of public affairs. He spent most of his time at Thoresby, improving and maintaining his estate out of an annual income of some £50,000 pounds, mainly from the Derbyshire collieries he owned. Between 1865 and 1875 the existing house was demolished and a new one, designed by Anthony Salvin (1799–1881) in a massive neo-Tudor style, was erected on a different site. A number of designers were consulted about the accompanying gardens. These included William Nesfield (1793–1881), Broderick Thomas (1811–98) and a 'Mr Inglis' but none of their proposals suited the earl. It seems that the final arrangement was the work of Salvin who produced a series of terraces in a typical late-nineteenth-century layout with carpet-bedding.

The view here is from the south-east with the main approach drive on the east. To the south, foreground, adjacent to the house, is a broad terrace with gravel paths. It formerly had elaborate summer bedding arrangements which have recently been re-established. From this terrace four flights of steps lead down to a lawn cut into by stone-edged geometric beds and with gazebos in the lower corners. Below a grass slope is a smaller lawn with more stone-edged beds and with a central projection containing a fountain and lily pool. To the west of the house, upper left, are two sunken lawns. These are divided by an axial path and cut into by beds. Beneath is a third sunken lawn with a large projecting semicircle to the west. Although its planting is much simplified, the structure of this garden is still largely intact.

C. Aslet, 'Thoresby Hall', *Country Life* 165 (1979), pp. 2082-5; B. Elliott, *Victorian Gardens* (London: 1986), p. 166; K. Lemmon, 'Geometry in a woodland landscape', *Country Life* 180 (1986), pp. 876–8; neg. no. DW 31, 13 July 1949.

49 *Wimpole Hall, Cambridgeshire*

Wimpole, now a National Trust property, lies 13 kilometres south-west of Cambridge. Its great park is notable for having been worked on by almost every major English landscape designer including Bridgeman, Greening, Brown, Emes and Repton. Here a more recent part of the Wimpole landscape, the traces of a nineteenth-century parterre, is visible.

The landscaping of Wimpole park in the second half of the eighteenth century led to the removal of the old formal gardens around the house and the extending of the park to its walls. In 1801 Repton advocated recreating a garden, set between the two north-projecting wings of the house, here seen in the photograph taken from the north. This proposal does seem to have been taken up but in a much more formal way than Repton would have wanted. By 1808 there was a garden here but it included a series of paths radiating from the central door.

The Repton alterations to the park, and presumably this garden, were commissioned by Philip Yorke, third earl of Hardwicke (1757–1834), whose grandfather, the first earl, had bought Wimpole in 1741. On the death of the third earl Wimpole passed to Charles Yorke, the fourth earl (1799–1873), and it is likely that the complex parterres, the remains of which are visible in the photograph, were created during his tenure. Neither their exact date nor their designer are known. They may be as early as 1850 and certainly by the end of the century they were being simplified. They were abandoned in the 1930s.

The light dusting of April snow picks out the slight remains and shows that each of the two northern parterres were divided into quadrants by broad cross-paths. In the centre of each quadrant was a yew or a cypress, some of which survived until the 1950s. These were approached by eight radial paths. The parterre to the south-west of the house was similar but less elaborate. Recent excavations by the National Trust have shown that all these parterres were constructed with imported soil, set in 30-centimetre deep beds. The paths were of gravel. Modern parterres, based on the original layout, have now been planted on the northern side of the house.

National Trust, *Wimpole Hall* (London: 1986); D. R. Wilson, 'Old gardens from the air', in A. E. Brown (ed.), *Garden Archaeology*, Council for British Archaeology Research Report 78 (1991), p. 33; neg. no. BSG 43, 9 April 1975.

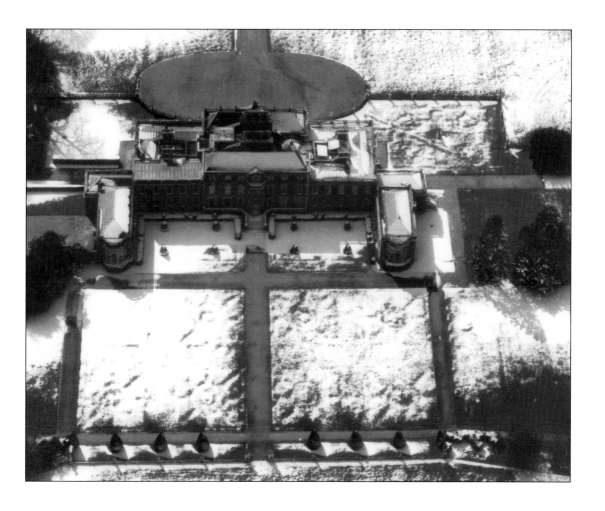

50 Penrhyn Castle, Caernarfonshire

Penrhyn overlooks the north-eastern end of the Menai Straits. The site is an ancient one, the medieval castle there being turned into a country house in the sixteenth-century. In 1765 the estate passed to Richard Pennant (1737–1808), the son of a Liverpool merchant with extensive West Indian properties. Pennant expanded his family's commercial interests by developing the slate quarries at nearby Bethesda, the products of which eventually provided roofs for much of Victorian Britain. Pennant was MP for Liverpool and in 1783 was created Lord Penrhyn. He rebuilt the house here but although the extent of his interest in gardening is unknown it seems likely that he was at least partly responsible for the great landscaped park. Pennant was succeeded by his cousin Henry Dawkins (1764–1840) who took the family name. It was Henry Pennant who, in the 1820s, engaged Thomas Hopper (1776–1856) to transform the house into a vast Norman-revival castle. This transformation was not completed until 1846 by which time the estate had passed to Pennant's son-in-law Edward Douglas (1800–86) who was created Lord Penrhyn in 1866. The estate remained in the hands of the family until 1949 and passed to the National Trust in 1951.

The castle, seen here from the south-west, is set on high ground dominating the surrounding park and seashore. The park sweeps up to the walls of the castle in a manner unusual for its time. The park was enlarged and enhanced by Edward Douglas-Pennant and his son George (1836–1907) by the planting of exotic trees from America and Australasia, many of which are visible in the photograph. West of the castle, in the left middle distance, is a detached, walled, formal garden, terraced into the slope and protected by woodland. It was created in the 1840s by Henry Pennant. Originally it had elaborate arrangements of luxuriant sub-tropical planting set between the regular paths and a central basin, although it has subsequently been modified on at least two occasions.

The reasons for the existence and the exotic appearance of this and other north Welsh gardens lies in a combination of factors. The relatively mild climate was important as was the romantic scenery. Equally significant was the construction of the Holyhead Road in 1818 and the building of the Chester–Holyhead Railway in 1848. These opened up the area and allowed access to it by the newly rich industrialists and entrepreneurs from Lancashire and the Midlands.

R. Haslam, 'Penrhyn Castle', *Country Life* 181 (1987), pp. 108–13; C. Hussey, 'Penrhyn Castle', *Country Life* 118 (1955), pp. 140–3; G. S. Thomas, *Gardens of the National Trust* (London: 1979), pp. 194–6; neg. no. ML 28, 25 July 1953 .

51 Leighton Hall, Montgomeryshire

Leighton lies 3.5 kilometres south-east of Welshpool. The estate was bought in 1845 by Christopher Leyland and in 1847 given to his nephew, John Naylor (1813–89), as a wedding present, together with £100,000 to improve it. Naylor was a millionaire industrialist, partner in the Liverpool bank of Leyland and Bullen and one of the richest men in Victorian England. He was one of those not uncommon new landowners who were prepared to develop their estates regardless of cost or profitability. He introduced advanced technologies to his land, including gasworks and hydraulic rams, and planted extensive woodland including a grove of Californian redwoods. He also raised the first Leyland cypress in 1888, named after Thomas Leyland, founder of the bank and the source of Naylor's wealth.

Leighton Hall, designed by W. H. Gee in 1850–6, is a vast Gothic pile giving views across a park to the church, also by Gee. Around the hall are the remains of Naylor's equally exuberant gardens which also reflect the confidence and idealism of their owner. They were created in the 1860s by Edward Kemp (1817–91) who was then designing increasingly complex arrangements.

Leighton exhibits this trend. The gardens are made up of a series of linked formal and informal set pieces. In the photograph, taken from the south, only parts of them are visible. In the right foreground, south-east of the house, is the principal compartment, a broad lawn below a terrace. Only remnants of the paths survive, but originally it had a complex parterre. The flower-beds, statuary and fountain have disappeared. Above, to the north and extending along the north of the house, is an informally planted area with a small lake on its east side. The small valley in which the lake lies extends southwards. It too was planted informally and is crossed by a bridge leading to the park. To the south-west of the hall is the formal walled Library Garden with its beds and paths. At its south end is a raised grass terrace with a gateway leading to a lawn and the woodland beyond. After Naylor's death his widow lived on at Leighton, the gardens remaining in impeccable order. On her death in 1909 the estate was broken up.

'Leighton Hall', *Country Life* 12 (1902), pp. 528–35; R. Haslam, 'Leighton Hall', *Country Life* 185 no. 26 (1991), pp. 116–19; C. Musson, *Wales from the Air* (London: 1992), pp. 154–5; neg. no. CJX 63, 25 July 1979.

52 Batsford Arboretum, Gloucestershire

Batsford lies 2.5 kilometres north-west of Moreton in Marsh, on land sloping south-east. An eighteenth-century house and park were bought in 1886–7 by A. B. Mitford (1837–1916), a diplomat who spent many years in the Far East and who became fascinated by Japanese culture. He was invalided home in 1870 and although he remained a civil servant – he was Secretary to the Board of Works from 1874 – he wrote widely on Oriental subjects. In 1886 his cousin the Earl of Redesdale died and left him a considerable fortune. Mitford retired to Batsford where he demolished the existing house and replaced it with a new one, here seen from the south-east, designed by Sir Ernest George (1839–1922) in a sixteenth-century style. It was completed in 1892. Mitford, who was created Lord Redesdale in 1902, then began the arboretum.

The arboretum soon contained a wide range of trees and shrubs from all over the world, perhaps partly as the result of advice from Sir Joseph Hooker (1817–1911) who was a friend of Redesdale. Because of Redesdale's Oriental interests, plants from there predominated and included maples, cherries, cedar and pine trees from China and Japan, and there was a Japanese Garden with an outstanding collection of bamboos. The planting was informal with trees and shrubs scattered over the entire area of some 20 hectares in varying densities. In addition, a number of buildings and other features generally of Oriental form were erected, including a Japanese 'rest-house', a rock garden, a grotto and some Japanese bronze statuary. The whole was created with considerable skill with open views across the country to the south-east.

In 1920 the estate was sold to Sir Gilbert Wills (1880–1956), later first Lord Dulverton, and second son of Sir Frederick Wills of the Bristol tobacco firm. After 1956 the second Lord Dulverton (1915–92) extended the original planting with many more exotic trees. An illustration of their variety can be seen in the curving line of trees in the centre background of the photograph. It contains two forms of Himalayan spruce, another from China and five different oaks from the Mediterranean, the Far East and North America.

'Batsford Park', *Country Life* 14 (1903), pp. 18–28; *Guide to Batsford Arboretum* (Batsford: 1992); J. Sales, *West Country Gardens* (London: 1980), p. 38; neg. no. YT 50, 22 June 1959.

Chapter 8

Twentieth-century Gardens

The Edwardian period, 1901–14, saw what was virtually the final flowering of great private gardens.[1] A number of major gardens did come into existence after the First World War and some of these, Sissinghurst **[53]** and Hidcote, Gloucestershire, for example, later became very influential; but, in tune with the great political, economic and social changes after 1918, gardens tended to become smaller, more intimate and, most notable of all, reached for the first time into every social class. After 1945, the larger designed landscapes that appeared tended to be associated with major constructional work for public and institutional bodies at both local and national levels. These included the landscapes of the various New Towns, here illustrated by Cumbernauld **[69]**. The settings of places like Oldbury Nuclear Power Station, Gloucestershire, and the Rutherford Appleton Laboratory at Harwell, Oxfordshire, both by Sir Geoffrey Jellicoe (1900–) in the 1960s, were others.[2]

Many of these landscapes, as well as the larger modern gardens, have been discussed and admired by professional gardeners and landscape designers, but they have had relatively little impact upon the great mass of small gardens of the twentieth century.[3] The owners and creators of these small gardens have studiously avoided much of the mainstream intellectual and philosophical base of garden design and have made gardens inspired by other sources. Thus, although the Modern Movement produced in the 1930s a number of notable and exciting gardens such as that at St Ann's Hill, Chertsey, Surrey, by Christopher Tunnard (1910–79), it had little or no effect on the mass of new gardens.[4] Nor did many of the important early-twentieth-century gardens, created in the late Victorian tradition. The continuing interest in the Italian style can be seen most notably perhaps at Hever [54] made by an Italophile, and probably more historically correct than anything earlier. Others include Sir Harold Peto's (1854–1933) own garden at Iford, Wiltshire, and later, Port Meirion, Meirionethshire, by Sir Clough Williams-Ellis (1883–1978) and Port Lympne [55].[5] Another major Edwardian garden, Mellerstain [56], is also of interest in this respect in that, among many others, it was laid out in a restrained, symmetrical, neo-Georgian style. This style dominated public and institutional architecture until the 1960s but again was never embraced by the increasingly affluent middle classes for either their homes or their gardens.[6]

The style which was popular for both houses and gardens from the late nineteenth century was the English Vernacular. This was in part the result of the influence of a number of architects, notably Sir Edwin Lutyens (1869–1949) and C. F. A. Voysey (1857–1941), whose works before 1914 were widely copied by other professional architects and by speculative builders.[7] The houses they designed were admirably suited to the equally vernacular forms of planting advocated by both William Robinson and Gertrude Jekyll. After the First World War both this architecture and this form of garden design caught the mood of the time, particularly as the 'suburban' expansion began. From then on every house had its garden. The houses tended to be a mixture of debased vernacular or 'applied' Tudor style and, at an almost standard density of 28 per hectare, afforded relatively spacious plots, usually unimaginative rectangles between 25 and 60 metres long and 6 to 15 metres wide. These produced small, squarish, front gardens and long, narrow, rear ones.[8] Formal Italianate terraces, balustrading or intricate bedding had no place here, but wide herbaceous side borders, small lawns and crazy-paved paths did.

Although the results of this expansion of new homes and gardens are visible around every town in Britain, it is difficult to appreciate its true scale. During the 1920s, 150,000 houses a year were being built in Britain, three-quarters of them by private builders. In the 1930s the numbers were even greater. In the worst year of the Depression, 200,000 houses were constructed, rising to 350,000 in 1936. By 1939 one-third of all houses in Britain had been built in the previous twenty years.[9] Thus, in these twenty years, over four million new gardens were created.

Few of these gardens can be regarded as of particular quality but they laid the foundations for even more gardens that appeared after the Second World War. By the early 1960s almost 400,000 dwellings a year were being built and although substantially fewer were erected in the next two decades there were never less than 200,000 per annum. In total, therefore, another eight million or so gardens appeared. Alongside these new gardens and their houses went a new social order, new values, new information, new technology and much, much more money. From the 1950s onwards it appears that, superficially at least, there was even more variety than ever in these millions of gardens as pools, patios, conservatories and barbecues arrived. These seemed to be especially prevalent in the New Towns such as Milton Keynes [57], Harlow, Stevenage or Basildon but can be seen in almost every town, suburb and village. Nor was the seemingly endless variety of gardens the result only of structural innovations or additions. The numbers and forms of available plants were revolutionised by developments in plant-breeding. New equipment, from strimmers to propagators, and new chemicals, from weed-killers to rooting-powder, made gardening easy for all. The flood of gardening programmes on television and on the radio, of gardening magazines and books, as well as the appearance of the now ubiquitous garden centre, made it theoretically possible for all gardens to be different.

Yet, despite these opportunities to create almost anything and while accepting the fact that passionate, individual gardeners do indeed exist, gardens on the whole at the end of the twentieth century have an underlying uniformity that verges on the dull and is certainly very repetitious. The historian of the modern landscape might find this both curious and difficult to explain. The answer is complex and lies partly in the continuing affection for the 'vernacular' traditions of Robinson and Jekyll. These were further developed, although still within firmly Jekyllesque limits, by other garden designers such as Norah Lindsey (1866–1948), Lanning Roper (1912–83) and Penelope Hobhouse (1929–). These designers advocated, and indeed created, gardens that touched the chords of perceived past gentility and romanticism but which could be adapted and modified cheaply and easily to suit suburban gardens. Equally important, especially since the 1960s and the all-pervading car culture, has been the 'National Trust effect' which has resulted in the standardisation of many of the gardens owned and acquired by the Trust and, to a lesser extent, by the National Trust for Scotland. The inevitable, broad, herbaceous borders in predominantly pastel shades so favoured by the Trust, have been carried back from Sunday outings to millions of suburban gardens.

Perhaps the greatest influence on modern gardens is the mid-nineteenth-century love of compartments, which developed out of a much older tradition. In the larger gardens of the later nineteenth and early twentieth centuries compartmentalisation was a device that enabled their owners to create 'themed' sections based on plant types, origins, forms or colours, or on historical or architectural set pieces. It was thus ideally suited to gardeners with interests in both plants and design. The compartmentalised garden at Dyffryn [58] shows both these interests, as does Great Dixter [59]. At the latter in particular both the ancient house and the compartmentalised garden produce for many both the ideal and idealised

English country setting. Even Hever **[54]**, although on a grand scale with its Italianate layout and its more intimate English arrangements close to the castle, is similar in inspiration.

But compartments also suited the new gardening democracy of Britain, especially after 1945. Any garden in the expanding suburbs or new towns could easily be given two or more compartments, each filled with beds, patios, lawns or shrubberies and linked by contrived views through trellis-work or arches. And two major modern gardens have helped more than any to foster this development, Hidcote in Gloucestershire and Sissinghurst **[53]**. Both comprise a series of compartments but, although Sissinghurst was planned thus from the beginning, Hidcote appears to have been the result of the steady accretion of compartments by its owner and creator Lawrence Johnston (1871–1958), between 1907 and the 1930s.[10] At first, both gardens, through the social contacts of their owners, influenced a number of other gardens including those at Newby Hall, Yorkshire, Tintinhull, Somerset and Kiftsgate, adjacent to Hidcote.[11] But since the 1950s, when both Sissinghurst and Hidcote passed to the National Trust, they have been visited by millions of people and have become the most popular gardens in the country. As a result, their impact has been enormous and there can hardly be a serious gardener in Britain who has not, consciously or unconsciously, been affected by them.

Two other influences on twentieth-century gardens need to be noted. The first is that of the so-called plantsmen's gardens, the owners of which collectively organised, paid for and took part in plant-hunting expeditions all over the world, and propagated, raised, hybridised and publicised a vast range of flowers, shrubs and trees. Some of these gardens are only marginally 'designed', if at all, as is the case at Caerhays **[60]**. Others, however, including Dyffryn **[58]** and Tresco on the Isles of Scilly, and also the group of Sussex gardens which includes Nymans, Wakehurst Place, Leonardslee and Borde Hill, are beautifully designed gardens in their own right.[12]

The other great influence of recent times has been the creation of 'historic' gardens using designs and details from the past, expressed in different ways and with varying degrees of historical accuracy. Hever **[54]** is an early example of 'correct' Italianate design, the later Port Lympne **[55]** a much less rigorous classical essay. Modern examples more firmly based historically are the medieval gardens at Southampton and Winchester, the sixteenth-century one at Hatfield House, Hertfordshire and the seventeenth-century garden at Pitmedden **[61]**.[13] There has also been a fashion for restoring or recreating historic gardens. Some of the earlier attempts, such as that by the then Office of Works at Kirby Hall, Northamptonshire, in the 1930s, now removed, have been derided. With the aid of expert scholarship from the Garden History Society and from English Heritage, modern restorations appear to be more historically correct than ever. Some notable successes are the late-seventeenth-century Privy

Garden at Hampton Court and the 1830s layout at Audley End, Essex.[14] Yet whether these restorations, or recreations, will stand the test of time remains to be seen. As with historically-based films, such gardens seem splendid when new but then increasingly declare their true age.

For obvious reasons, the motives behind the creation of twentieth-century gardens seem clearer than those of any previous time. The expert and passionate plantsmen are well known and their gardens survive, as at Caerhays **[60]**. Likewise, there is no doubt about the interest of the Lloyds of Great Dixter **[59]**, Cory at Dyffryn **[58]** or the Nicholsons at Sissinghurst **[53]**. Nor, at the other extreme, can Port Lympne **[55]** be seen as anything other than an elaborate stage set for Sassoon's political and social gatherings. Hever **[54]** is perhaps more complex. Astor's interest in gardening is well known, but as an American who became a British subject he also needed to bolster his status and required an outdoor display case for his classical antiquities. Not enough is known of Portman to be sure about Hestercombe **[62]** although the dichotomy between an architecturally unsatisfactory house and the superb garden is curious. Likewise, Mellerstain **[56]** was apparently designed to be a low-maintenance garden despite its spectacular appearance and its views. What lay behind it is by no means clear.

The motives behind the millions of small gardens in suburbs, in villages, in New Towns **[57]** and in ancient cities, can also only be guessed at, although they are probably as wide-ranging as such motives have always been. Only the vast numbers are different, making the landscape more varied than ever before and, for the landscape historian, more interesting.

53 *Sissinghurst Castle, Kent*

Sissinghurst is perhaps not as impressive from the air as many other gardens. However, it is arguably one of the most influential of the twentieth century. It was created between 1930 and 1950 by Vita Sackville-West and her husband Harold Nicholson. Sackville-West (1892–1962), was a poet, novelist, and later a much admired garden writer. Her father was Lionel, third Baron Sackville of Knole. Harold Nicholson (1886–1968) was a diplomat and a son of Arthur Nicholson, also a diplomat and later Lord Carnock.

Sissinghurst was originally the site of a medieval moated manor house, rebuilt between 1535 and 1558. When the Nicholsons purchased it in 1930 it was derelict and only the great gate tower, some outbuildings and sixteenth-century walling survived. The walls were used as a framework within which a series of enclosed gardens were laid out. Probably due to Harold Nicholson, the enclosures were cleverly linked into a coherent whole by long axes. The planting was the work of Vita Sackville-West in the tradition of Gertrude Jekyll. The result is an assymetrical arrangement, formal in layout but planted informally. As John Sales has said, it is 'the epitome of the English Garden, intimate in scale and rich in plants . . . given cohesion by . . . the unifying effect of a strong architectural framework'.

The photograph is from the north-west. In the foreground is a sixteenth-century range of former outbuildings. Behind them is the broad grassed Tower Courtyard (1931–2) edged by wide herbaceous borders. To the north, left, is a small compartment called Delos (c. 1940) with informal planting. Above it is the White Garden (1946) with grey and white shrubs, trees and flowers, set in a formal arrangement of paths. To the south, right, of the Tower Courtyard the Rondel Rose Garden (1933–4) is visible, with its main east–west axis and with the long Lime Walk (1932–3) set at an angle on its south. Beyond, to the east, is the Nuttery, with hazel trees underplanted with spring flowers and with the formal Herb Garden, view beyond. In front of this is the Moat Walk (1930), a sunken lawn within a the medieval moat.

The extensive wooded area in the north-east is an orchard, bounded on the nor by the arms of the still water-filled moat. To the east of the Nuttery and the north of Walk is the South Cottage with its formally arranged but informally planted garde

J. Brown, *Sissinghurst, Portrait of a Garden* (London: 1990); G. Jellicoe (ed.), *Oxford Companion to Gardens* (Oxford: p. 519; V. Sackville-West, *The Garden* (London: 1946); neg. no. BIJ 55, 23 May 1972.

55 Port Lympne, Kent

The gardens at Port Lympne lie just below the crest of a steep unstable hillside, in fact an old line of sea cliffs, looking out to sea. The house was designed in 1912 by Sir Herbert Baker (1862–1946) in a Cape Dutch style for the millionaire Sir Philip Sassoon (1885–1939), only son of Sir Edward Sassoon and Alice de Rothschild. Educated at Eton and Oxford, Sir Philip consciously chose British nationality at the age of 19. He was MP for Hythe from 1912. During the First World War he was private secretary to Field-Marshal Haig and in 1920 he held the same post to Lloyd George. In 1924–9 and 1931–7 he was Under-Secretary for Air. In addition to his political duties Sir Philip was a great society host in the 1920s and 1930s and as a result, most of the political, literary and artistic personalities of the period enjoyed lavish hospitality at Port Lympne. In the early 1920s the house was used for meetings associated with the peace conferences and international statesmen were regular visitors.

As Jane Brown has written, the gardens 'were not a labour of love, nor of sensitive harmony with nature. They were the flamboyant imposition of one man's pride and wealth'. Like the house, they were intended to impress and to provide a setting for social events. They were conceived by Sir Philip himself but largely designed by Philip Tilden (1887–1956) in the early 1920s. Some remodelling was carried out by Russell Page (1906–85) in about 1974, after this photograph was taken. In their heyday these gardens required a staff of thirty to keep them in perfect order.

The view here is from the south across the hillside. To the west, left, of the house is a formal water garden from which rises the 135-step Trojan Stairway, flanked by clipped hedges and terminating in a circular viewing platform. To the west again is a long east–west avenue and below it a flat area, once the bowling green, where marquees were erected in the Season. On the south side of the house, between curved sitting-out places, is a balustraded terrace. Below it at the west end is the Chess Garden, enclosed by clipped yews and chequered with annuals, while at the east is the similar Striped Garden. Between them is now a simple pool which, until the late 1920s, was the site of three lavishly decorated neo-Roman swimming pools, each with a fountain and classical stone seats. Serious subsidence led to their demolition.

Below the pool the garden falls away to more terraces, a vineyard and fig-yard as well as two tennis courts, one for morning and one for evening. Between the central section and the Bowling Green is a long sloping path separating what are now herbaceous borders. It is matched on the east by a similar one broken by a stone staircase.

J. Brown, *The English Garden in Our Time* (Woodbridge: 1986), pp. 95–9; G. C. Taylor, 'The gardens at Port Lympne', *Country Life* 66 (1929), pp. 513–17; G. C. Taylor, 'The gardens at Port Lymphe', *Country Life* 79 (1936), pp. 276–82; neg. no. QM 27, 2 July 1955.

56 Mellerstain, Berwickshire

Mellerstain lies nine kilometres north-west of Kelso. In 1725 William Adam (1689–1748) added wings to an existing house for George Baille (1663–1738). The gardens, including a long canal, were then apparently in a formal Dutch style, probably as a result of the exile in The Netherlands in the 1650s of Baille's royalist father. In 1719 one of Baille's daughters married Charles Hamilton, Lord Binney (1697–1726), eldest son of the sixth earl of Haddington, and thus Mellerstain passed to the Hamiltons. The seventh earl completed the rebuilding of the house in 1770–8 and may then have replaced the gardens by parkland although the earlier canal remained.

Few changes were made until 1909 when Lord Binney (1856–1917), eldest son of the eleventh earl, having retired after a distinguished army career, commissioned Sir Reginald Blomfield (1856–1942) to design a formal but manageable garden. Blomfield was primarily an architect but he was also a proponent of structural symmetrical formality for gardens, which he believed should extend across the landscape providing long vistas. Mellerstain is the epitome of this concept and presumably it was Blomfield rather than his client who was the overriding influence on its design, although the gardens were ultimately not as extensive as Blomfield wished.

The view here is from the south. Immediately below the house is a formal terrace from which a double stairway descends to a walled compartment, containing two box-edged parterres separated by an axial path and with raised platforms at its lower end. In the centre of the lower side of this compartment another double stairway descends to a long lawn which is separated from the adjacent parkland by walling and inner hedges. The lawn is in two parts. The upper broad section is separated from the long narrow lower one by a grassy amphitheatre. At the bottom of the lawn is a stone-revetted crescentic walk above a lake. Blomfield created this lake by widening the old canal and reducing its formality by giving it sinuous edges. The view across the lawns and lake extends to the Cheviot Hills, 25 kilometres away. To the north-west of the house the parkland was also formalised with long vistas through plantations.

B. Elliott, *Victorian Gardens* (London: 1986), p. 242; M. Girouard, 'Mellerstain', *Country Life* 124 (1958), pp. 416–19, 476–9, 1067–9; D. Ottewill, *The Edwardian Garden* (London: 1989), pp. 24–6; neg. no. GT 42, 16 July 1951.

57 Milton Keynes, Buckinghamshire

Not as visually impressive as Cumbernauld **[69]**, although more demanding of land, Milton Keynes represents the point which gardens and designed landscapes had reached at the end of the twentieth century. It is one of the third generation of post-war New Towns and was intended as a home for a car-owning community, but where the car was separated from the people. Its overall plan is based on a loose one kilometre-square grid of primary roads within which housing, industry, recreational facilities and amenities are located. Planning began in 1967 and building commenced in 1970. It is now a town of almost 200,000 people. In this view, from the south, part of the more recent development at Far Bletchley is visible.

Milton Keynes is an experiment in social engineering set in the largest designed landscape in Britain, but with its low housing density, which enables almost every home to have a garden of some kind, it also represents the democratisation of gardening. The multitude of gardens here exhibit more extensively than ever before the same variety of human response to gardening as in every previous age. Some gardens are carefully designed and laid out. Others are simply crammed with plants. Some are intended to impress the neighbours and others are the result of a passion for gardening. Some are given minimal attention, others are cared for lovingly.

All are the result of the interweaving of two threads. One is the whole history of garden and landscape design, plant collecting and breeding, which this book attempts to illustrate. The other is the more recent social development whereby, for the first time, almost everyone can have a garden. But these gardens also reflect the curious contradiction inherent in human nature. At the same time as the millions of these gardens are created and maintained, often at great expense, their owners are prepared to allow other, greater, gardens and landscapes to be ploughed flat to produce cheap food as at Eastbury **[78]** and Stainfield **[79]**, to be carved up by motorways as at Tredegar **[74]** or to be built over as has happened at Great Barton **[76]**. And, in a wider context, to see natural wetlands emptied of peat, plants stripped from their natural habitats, and forests felled.

Neg. no. RC8-FT 138, 28 July 1983 .

58 Dyffryn, Glamorgan

Dyffryn lies five kilometres north-east of Barry. A house and small park existed here from the eighteenth century but little is known of either. In 1891 the estate was bought by Sir John Cory (1828–1906), philanthropist and member of a wealthy family of colliery owners and shipping magnates. The family originated in Devon but moved to South Wales and eventually founded a group of companies, all involved in the mining and transport of coal. Much of this coal was exported through Barry docks, the development of which the Corys promoted.

Sir John had the house at Dyffryn rebuilt in 1893–4. He also laid out two balustraded terraces to the south and a formal garden to the east. These were retained when, in 1903–4, he commissioned Thomas Mawson (1861–1933) to lay out a more extensive garden. On Sir John's death Mawson continued to work for and with his son Reginald (1871–1934). Reginald Cory was a notable plantsman who paid for and went on plant-hunting expeditions and who eventually had outstanding collections of trees, shrubs, dahlias and water lilies. He was also interested in garden design and worked closely with Mawson. The result of their collaboration was a remarkable garden with intricate formal and informal compartments.

The view here is from the south-west and only part of the extensive layout is visible. Immediately below the house are the two original terraces, both bounded on the south by yews. Below is the Great Lawn, itself bounded east and west by grass slopes. The lawn has an axial canal with a rectangular central pool and an octagonal southern one. At the south end of the lawn is a pergola flanked by open pavilions below which is a court with areas of formal and informal lawns and shrubs. At the south-west corner of the lawn is the Pool Garden, the paths and curving lily pools of which are just visible.

To the west, left, of the Lawn is a group of linked compartments and a larger informal area of trees and shrubs. Of the former, the circular Topiary Garden with radiating box-edged beds and paths and with the Paved Court above it, half-left, are the clearest. The latter is stone-paved with inset beds. To its right is the Pompeian Garden with an arrangement of colonnaded loggias and fountains. West of the Paved Court is the Bathing Pool Garden and above it, half-left, is the Theatre Garden which originally held Cory's bonsai collection. Further compartments lie to the north. Cory continued to develop the gardens until 1931 when he retired to Dorset. On his death they were sold. They are now in the hands of the Vale of Glamorgan Council.

T. Mawson, *The Art and Craft of Garden Making* (London: 1926), p. 386; J. Newman, *Glamorgan* (London: 1995), pp. 341–2; neg. no. CBX 25, 4 March 1977.

59 Great Dixter, Sussex

A very influential garden although, perhaps, as much through of the writings of its present owner than because of its design. The house, at Northiam, 15 kilometres north of Hastings, was built in the 1460s. It was remodelled and enlarged for Nathaniel Lloyd (1867–1933) when he purchased the property in 1910. The architect was Edwin Lutyens who also designed the garden.

Lloyd was originally a printer and later an architectural historian of some note. He was also a skilled gardener and was especially knowledgeable about topiary. The garden was laid out as a series of enclosed areas, bounded by hedges and walls and linked by steps and doorways. Yet, apart from the terrace at the rear of the house with its complicated circular steps, the underlying structure is muted, unlike that at Hestercombe [62] by the same designer. This was because Lloyd was a proponent of informality and thus subsequently softened Lutyens's architectural structure.

The view here is from the west. To the front, left, of the house is a yew-hedged wildflower meadow. Below it, to the north-west and bounded on two sides by outbuildings, is the Sunken Garden. This was an addition to the original plan by Lloyd himself in 1923. It has a central octagonal pool below low terraces. To its south, right, is a small walled compartment and right again a broad lawn with free-standing clipped yews. Above this is another yew-hedged enclosure with an old cattle byre on the west side, altered to form a rustic loggia. In the photograph the interior is a formal rose garden although this has since been made more informal. The large open south-eastern section is an orchard and another more extensive wild meadow, while the north-eastern part is yet one more hedged compartment with a kitchen garden beyond. Immediately to the right, south, of the house is a formal terrace which extends east and becomes the 60-metre Long Border, planted with shrubs, herbaceous plants and annuals.

After Lloyd's death the garden was taken over by his son, Christopher Lloyd (1921–), a renowned plantsman and writer. His interest in luxuriant planting has continued to reduce the impact of the formal structure of the garden.

C. Lloyd, *The Well-Tempered Garden* (London: 1970); C. Lloyd, *The Year at Great Dixter* (London: 1987); L. Weaver, 'Great Dixter', *Country Life* 33 (1913), pp. 18–26; T. Wright, *The Gardens of Britain* 4 (London: 1978), pp. 144–50; neg. no. CPN 15, 8 June 1982.

60 Caerhays, Cornwall

Caerhays is one of the many gardens that benefit from Cornwall's mild climate. It lies 12 kilometres south-south-west of St Austell, overlooking Porthluney Cove. The house, a mock castle designed by John Nash in 1805–7, was on a new site. Its owner, J. B. Trevanion (1780–1840), was a member of an old Cornish family who had held Caerhays for centuries. Trevanion also laid out a landscape park about which Humphrey Repton is said to have been consulted in 1809. While this has not been proved, a long serpentine lake was created then, visible within the parkland at the top of the photograph, here taken from the south-west.

The primary interest of Caerhays, however, is as a plantsman's garden. After Trevanion had dissipated the family fortune, Caerhays was acquired in 1853 by J. M. Williams, a member of a family made rich by their mining interests. The estate was changed little until 1896 when Williams's second son, John Charles Williams (1861–1939), retired to Caerhays after a brief spell as MP for Truro and spent the rest of his life there.

Williams was a plant collector and hybridiser and spent considerable sums sponsoring plant-hunting expeditions. These were mainly to Burma, China, India, Japan and Tibet and included ones by G. Forrest (1873–1922), R. Farrer (1880–1920), F. Kingdon-Ward (1885–1958) and especially by E. H. Wilson (1876–1932). Wilson's expeditions were also sponsored by Sir Harry Veitch (1840–1924), the last head of the famous firm of nurserymen, who ensured that many of Wilson's seeds went to Caerhays. The often delicate new plants and seeds arrived at irregular intervals over many years. As a result, there was never any overall design or structure. The gardens were gradually expanded as areas within the existing woodland were cleared, as at the bottom right.

Most of the development took place between 1903 and 1926 but it was not until Williams's death that it and the extensive work of hybridisation stopped. Since then the gardens, which are still owned by the Williams family, have improved as many of the exotic species have reached maturity. These include notable varieties of rhododendron, azalea, magnolia, camellia and hydrangea.

N. Kingsbury, 'Woodland wonder', *Country Life* 188 (1994), pp. 62–5; S. Pring (ed.), *Glorious Gardens of Cornwall* (Oxford: 1996), pp. 57–8; J. Williams, 'The decline and rise of Caerhays', *The Garden* 117 (1992), pp. 62–4; neg. no. MF 82, 4 July 1953.

61 *Pitmedden, Aberdeenshire*

Pitmedden lies 22 kilometres west of Aberdeen. The spectacular design and planting is modern and its structure is of limited historical significance. In 1603 the late-medieval fortified tower house here became the home of the Seton family. When Sir Alexander Seton (c. 1630–1719), whose father had been killed fighting for his king in 1639, succeeded to Pitmedden on the death of his brother in 1667, the old Highland way of life was ending. Seton was a lawyer and an MP and was created baronet in 1683. He began the garden here around 1675, the date inscribed on a door lintel. No designer is known although it has been suggested that the architect Sir William Bruce (c. 1630–1710), a friend of Seton, was involved.

What remains of this garden, here seen from the east, is a broad terrace in front of the house with central formal steps leading down to the main enclosure. This is rectangular and is bounded by terraces backed by walls on three sides and by the upper terrace on the west. Two stone pavilions in the south-west and north-west corners are modern.

The Seton family retained Pitmedden until the late nineteenth century. The house was pulled down and replaced by an undistinguished structure in 1860 by which time a kitchen garden occupied the former garden. After Sir James Seton (1835–84), the eighth baronet, committed suicide, his brother Sir William (1837–1914), a soldier, sold Pitmedden to Alexander Keith in 1894. His son, Major James Keith (1879–1953), gave Pitmedden to the National Trust for Scotland in 1952. The then derelict garden was recreated. As no original planting scheme was known, four new box-edged parterres were laid out with modern bedding plants. Three were modelled on ones in the 1647 plans of Holyrood Palace, while the north-western one displays the arms of Sir Alexander Seton.

'Three hundred years of a great garden', *Country Life* 157 (1975), p. 1187; A. G. Hellyer, 'Scotland's Renaissance gardens', *Country Life* 148 (1970), pp. 402–4; G. A. Little, *Scotland's Gardens* (Edinburgh: 1981), pp. 206–8; neg. no. BVE 64, 26 July 1975.

62 *Hestercombe, Somerset*

Hestercombe lies four kilometres north-north-east of Taunton and occupies the site of an older house and its eighteenth-century park. The present house was built in 1874–7 in an unremarkable Renaissance style for Edward William Portman (1856–1911), the eldest son of the second Viscount Portman. The family had extensive estates in Dorset, Somerset and Central London which produced great wealth. Edward Portman built Hestercombe for himself when he was under twenty, which may explain the architecture. He lived there until his death for, as his father did not die until 1919, he never succeeded to the title or to the main estates.

The garden was begun by Portman in 1906 although he did not live long to enjoy it, for it was not completed until 1910. It was designed by Edwin Lutyens and Gertrude Jekyll and was the first garden where the two had almost complete freedom. The result is a masterpiece, a set of open-air rooms on different levels full of Lutyens devices and Jekyll planting, yet unified into a geometric whole.

The view here is from the south-west. The plain upper terrace is of the 1880s. Below lies Lutyens's Great Plat, a sunken parterre, the planting of which is divided diagonally by broad grass walks edged by flagstones. The Plat is bounded by stone walls supporting terraces and on each side, reached by sets of steps, are long narrow water gardens. Each of the last has an axial stone-lined rill, over 130 metres long, broken at regular intervals by pairs of circular pools, a Lutyens speciality. On the south side of the Plat, in the foreground, is a stone-pillared pergola 72 metres long. On the north-west, left, of the Plat broad steps lead to a rectangular rose garden which is bisected by the upper part of the eastern rill. In the north-east corner is The Rotunda, a circular walled and paved court with a central pool, which acts as a pivot for the whole garden. From it, set at an angle above a flight of steps, runs another stone-flagged terrace on which stands Lutyens's classical orangery. Beyond is a Dutch-style rose garden, unrestored when this photograph was taken.

J. Brown, *Gardens of a Golden Afternoon* (London: 1982), pp. 81–5; G. Jellicoe, *The Oxford Companion to Gardens* (Oxford: 1991), p. 254; neg. no. MH 94, 5 September 1953.

Chapter 9

Public and Institutional Parks and Gardens

Public and institutional gardens are different from private ones and have their own separate history. Public parks and gardens are those that were designed as, or have become, places open to the general public for recreation. Institutional parks and gardens are those that surround schools, hospitals, factories and other commercial buildings and are used primarily by the people occupying those places.

Public parks have a long history. The first examples appeared in the seventeenth century, all over Europe, and were then usually merely walks where the upper echelons of society met, talked and promenaded, largely for social reasons. In Britain one of the earliest such places was Moorfields in the City of London where, in the early seventeenth century, tree-lined walks were created.[1] Such walks became very popular in the eighteenth and early nineteenth centuries and many towns acquired them. At Dorchester, Dorset, the old Roman walls along the south and west of the town were removed in the late eighteenth century and replaced by avenues along which fashionable folk perambulated. At Leicester too the New Walk was laid out south of the town in 1785.[2]

From the seventeenth century, some of the London Royal Parks, which until then had been private Crown domains, gradually acquired a tradition of public access, but again mainly for the upper classes. This was, in part, the result of the late-seventeenth-century and eighteenth-century expansion of London that resulted in a lack of open spaces and in the formerly rural Royal Parks being surrounded by housing. For the urban rich, and especially those in the increasingly popular spa towns, this lack of open spaces was also ameliorated by the separate development of pleasure grounds, although these were open only to subscribers, and by squares, circuses and crescents that were part of the new developments in town planning. The squares and crescents usually had central gardens, which again were open only to the inhabitants of the adjacent properties. The first of these, although not strictly a garden at all, was Covent Garden in London of the 1630s. True gardens, as at Bloomsbury Square in 1661 and St James's Square in 1665, soon followed.[3] Thereafter, squares and crescents, all with private gardens, became common and they still form a major part of the urban landscapes in the West End of London and in Edinburgh [64], as well as in Bath, Brighton, Buxton, Cheltenham and other early resorts and spas. The greatest of these 'private' gardens was Regent's Park in London, 1811–26. Its origins lay in a speculative residential development by the Crown and only the inhabitants of the surrounding houses were allowed access. It finally became a public park in 1838 but before that it inspired a number of smaller examples, the most notable of which is Calverley Park [63].[4]

In the early nineteenth century there was also a growing demand for true public parks, led particularly by the garden designer and writer J. C. Loudon (1783–1843). The movement stemmed from the increasing awareness of the nature of the new industrial towns and thus of the need for public open spaces, coupled with the belief that parks could be instruments for social reform and education. The first public park was Derby Arboretum of 1840 which was intended as much for education as for recreation. The first truly recreational one was Birkenhead Park of 1843. The greatest on both counts was the Crystal Palace Park at Sydenham, London, of 1854. The early parks were characterised by generous planting, especially of exotic trees, and winding paths. As the century advanced and more parks were created, many features of fashionable private gardens, such as massed bedding and three-dimensional sculptural bedding on wire frames, were taken up. These features remained popular for much longer than in private gardens and indeed some survive to the present. With the appearance of the new and more powerful local authorities in the late nineteenth century, the role of public parks began to change. They acquired features that were primarily for entertainment rather than for education. Bandstands, fountains, floral clocks [64] and refreshment pavilions [65] all appeared, as well as provision for sports [63]. At first the areas for sport were limited and were easily fitted into the parks, but with changing social attitudes demand grew and spaces for cricket, football, tennis and bowls were required. In the 1920s and 1930s provision even for cycle-racing and athletics had to be made. As a result, the subtler aspects of planting were neglected.[5]

During the later nineteenth century the development of railways, which brought holidays and day-trips within the reach of all, led to the appearance of another type of public open space, the seaside garden. The railways produced a new urban form, the coastal resort. By the 1880s Britain was edged by holiday towns such as Broadstairs, Bournemouth, Weston-Super-Mare, Colwyn Bay, Cleethorpes, Skegness and Cromer.[6] All these places had public gardens characterised by elaborate flower-beds and opportunities for entertainment to draw in yet more people, as at Whitby [65]. Although not strictly public, but rather private and with commercial overtones, Thorpeness [66] is another seaside landscape. Its lake and islands are part of a middle-class resort village which could only be of early-twentieth-century date. It marks the beginning of a tradition of different resort landscapes which developed, through Butlin's and Pontins holiday camps, into Center Parcs.

Another form of public open space was that which followed from the increasing civic pride of both old and new Victorian towns. Town halls, art galleries, museums and law courts were often set around municipal squares which were eventually filled with fountains, statues of worthies, war memorials and colourful flower-beds. Cathays Park, Cardiff [67] is one of the last, and perhaps the best, example, while Princes Street Gardens, Edinburgh [64], is related but has a complicated earlier history.

In the second half of the twentieth century public parks have declined in status and appearance under the onslaught of yet more sports facilities, car parks and the demands of easy maintenance. But some, often quite obscure ones, have not only survived but, as at Lichfield [68], have managed to retain elements of their long history. Now public parks have been augmented by designed landscapes around housing estates and new towns such as Peterborough, Stevenage and Cumbernauld [69]. These, although often created by expert landscape architects and widely admired by the profession, tend to be repetitively dull, especially in their planting.[7]

Institutional parks and gardens have had a shorter history and often an even shorter life. Many institutions, particularly schools, universities and hospitals, but also modern factories and office blocks, were, and continue to be, set within designed landscapes. Yet because of the very nature of these institutions few of these landscapes remain intact. Expansion, technological development, social change and advances in medical care all mean that institutional landscapes are always under pressure from new buildings, car parks or playing fields. Such pressures can be seen at Christ's Hospital [70] and Exeter University [71].

The motives behind these institutional landscapes vary considerably. Sometimes they are part of a corporate or institutional image, sometimes they are meant to convey environmental commitment. At other places the gardens and landscapes are a genuine attempt to have better working or living conditions, as at Cumbernauld [69]. For hospitals and educational establishments they are for actual, or perceived, physical and mental health reasons. Many of the small General or Cottage Hospitals of the early twentieth century had gardens as part of their original designs. These were seen as good for patients and gratifying for the subscribers who had paid for them. These gardens usually had simple layouts with sinuous paths through lawns, herbaceous beds and shrubberies, and were protected or screened by belts of trees.

The purpose-built mental hospitals of the first half of the nineteenth century were given two types of landscape, secure inner courts, known as airing grounds, and open parkland beyond. As a result of developments in the methods of treating the mentally ill the airing courts gradually disappeared and were replaced by more open gardens of conventional type. The often extensive parkland beyond was always used for recreation and horticulture. These generous grounds were largely the result of the nineteenth-century Commissioners in Lunacy insisting on the provision of one acre (0.4 hectares) of land for every ten patients. Such parkland was usually planted extensively with trees but had few other special landscape features. Crichton Hospital [72] shows the open parkland effect as well as the more recent inner gardens that are now increasingly rare as a result of changes in the treatment of the mentally ill.[8]

Sanatoria, which were common from the later nineteenth century until the virtual elimination of tuberculosis after the Second World War, often had extensive grounds which included lawns, shrubberies and flower-beds near the hospital and outer grounds beyond, with woods and walkways. Most have now disappeared although those at the Royal National Hospital of 1869–75 at Ventnor on the Isle of Wight were distinguished enough to become the basis of the present Ventnor Botanic Gardens after the hospital was demolished in the 1970s.[9] By far the best surviving sanatorium garden is that of the King Edward VII Hospital of 1901–6 at Midhurst, Sussex. Its terraced gardens were added in 1906–7 to designs by Gertrude Jekyll who personally supervised the planting.[10]

From the late nineteenth century many of the new or relocated public schools, for example Christ's Hospital [70], as well as the then new redbrick universities, were also often given elaborate gardens or park-like surroundings in keeping with the ethos of the time. Birmingham University, for example, acquired an extensive well-treed campus after 1900. The newer universities such as Exeter [71], Keele, Sussex and East Anglia were all located in existing parkland. From the 1950s onwards even newly-built state schools had landscaped grounds, continuing the tradition of those at places such as the Cambridgeshire Village Colleges begun before the Second World War. However, all these institutions have suffered a slow attrition of their landscapes through inevitable expansion.

One final type of institutional landscape is the botanic garden, the origin of which lies in the beginnings of scientific enquiry in the Renaissance. The first was founded at Pisa in 1543, the first in Britain at Oxford in 1621. The Edinburgh Physic Garden (1670), The

Chelsea Physic Garden (1673) and Kew Gardens (1750) followed. The Cambridge Botanic Garden **[73]** was a latecomer, not begun on its present site until 1846. It exemplifies this type of garden in that, although primarily for academic research, it is open to the public and has a designed landscape.

63 Calverley Park, Tunbridge Wells, Kent

Although the medicinal value of the springs which led to the development of Tunbridge as a spa was discovered in the early seventeenth century, the town's success was long delayed. Visitors apparently stayed elsewhere and, although public buildings existed by the 1660s, it remained a fairly insignificant place for another century. It was still called 'a hamlet' in 1810. Then, in the 1820s, Tunbridge suddenly became fashionable and a period of rapid residential expansion began.

One of the earliest developments took place to the north-east of the town centre on some 20 hectares of land set around and in a steep-sided valley. Its owner was a John Ward (1770–1858) who in 1827–28 engaged Decimus Burton (1800–81) to design a garden suburb with its own private landscaped park. The plan was inspired by the Regent's Park development of 1811–26 by John Nash, with about one-third of the area developed for housing and the rest landscaped. It thus provided its middle-class residents with a designed landscape similar to that of a contemporary country house. The work was completed in 1840.

In the photograph, taken from the south-west, the buildings, all designed by Burton, can be seen arranged around the parkland. Clockwise from top left of centre they comprise the Calverley Hotel, originally a private house but enlarged and opened as a hotel in 1840; Calverley Park Crescent, a curved terrace of seventeen houses; and, below, Calverley Park, seventeen separate villas in a semicircle, all with gardens but all also sharing the private parkland. The original layout included a curved strip in front of the villas with shrubberies and meandering paths, although this is now mainly grass with mature trees. The valley below was originally an open grassy area with a lake and a wood at its lower end. Since about 1910 the valley has been developed as a public park by the local authority and has acquired the inevitable tennis court, bowling green, bandstand, herbaceous borders and a sunken rose garden.

N. Pevsner, *West Kent and The Weald* (London: 1969), pp. 553, 560–1; C. Hussey, 'Calverley Park', *Country Life* 145 (1969), pp. 1080–83, 1166–9; neg. no. AQA 74, 9 August 1966.

64 Princes Street Gardens, Edinburgh

The medieval town of Edinburgh lay mainly to the east of the great castle, which is visible in the foreground of this photograph, taken from the south. Immediately north of the castle was the marshy valley of the Nor' Loch which prevented urban expansion in that direction. The draining of this valley by the city fathers after 1759 and the acquisition of the land beyond was followed by the preparation in 1766 of plans by James Craig (1744–95) for a New Town on a gridiron plan. It was designed to consist of five parallel long streets, intersected by seven short ones. The outer streets, Princes Street on the south and Queen Street on the north, were to be built up on one side only in order to give views beyond. Work began in 1767 and the plots were sold to individual developers with limitations on what could be erected. As a result this first New Town lacks the overall architectural unity of similar contemporary developments, but it is still a remarkable piece of urban planning.

The valley below the castle is now Princes Street Gardens. After 1816 the area was developed initially as private gardens belonging to the owners of the adjacent properties across the street. Despite fierce opposition these gardens were mutilated by the construction of the North British Railway in 1844–6. In 1876, what remained was taken back into the hands of the city and a unified design created by Robert Morham (1839–1912). This gradually acquired all the features typical of late-Victorian municipal gardens, including extensive borders of spring and summer bedding plants, which still exist on the slopes below Princes Street. Under the direction of John McHattie (c. 1859–1923), the Superintendent of Edinburgh Parks, the first floral clock in Britain was created and it too survives in the north-east corner.

Since then the gardens have, inevitably, been simplified with areas of grass and shrubs. They have also been dignified by the erection of various regimental war memorials and other statuary including one of Thomas Guthrie, founder of the Ragged Schools. The more recent open air theatre in the centre is less attractive. The long strip of woodland in the background is the Queen Street Gardens, which separates the first New Town area from the second stage of development of 1801–26.

B. Elliott, *Victorian Gardens* (London: 1986), p. 211; J. Gifford, C. McWilliam and D. Walker, *Edinburgh* (London: 1984), pp. 271–314; H. Maxwell, *Scottish Gardens* (London: 1908), pp. 54–8; neg. no. AWB 70, 28 July 1968.

65 *West Cliff Gardens, Whitby, Yorkshire, North Riding*

The West Cliff Gardens are typical of many at seaside resorts. They lie on the west side of the old port of Whitby and are here seen from the north. Whitby has a long history but the photograph shows only its more recent past as a developing resort. The process was started by George Hudson (1800–71), the Victorian entrepreneur and 'Railway King'. He first promoted the Pickering–Whitby railway and then absorbed it into his railway empire in 1847. Despite his fall when the railway mania bubble burst in the same year, Hudson bought the West Cliff fields in 1848 and began to develop them for hotels and boarding houses. The Victoria Spa and Pump Room, erected in 1844 and visible on the cliff just right of centre, was the focus.

The earliest elements were the East Terrace of 1850, the L-shaped block on the upper left of the picture, and the zig-zag link-road to the harbour, still called the Khyber Pass. However, although the streets to the west were laid out, the development stalled until the 1870s when a new promoter, Sir George Elliot (?–1904), purchased the land and built up the grid of streets in the centre background. He also began the construction of the Royal Crescent in the centre middle distance. However, this was never completed and the western, right-hand, part of the planned crescent was finally occupied by the Metropole Hotel in 1897–8. Elliot was also responsible for the laying out of the crescent and cliff-top gardens as well as for the construction of the West Cliff Saloon, a reading room and refreshment bar, next to the Spa.

During the First World War the gardens and the undeveloped land were acquired by the Town Council who, in the 1920s, added an Orchestra Pavilion and laid out tennis courts, a bowling green and a putting green at the western end, right of the area. The formal gardens with their elaborate arrangements of annual bedding plants were re-established in 1946 after years of neglect. They have since been much simplified and partly abandoned. Together with the buildings they illustrate the failure of Whitby to become a major resort.

G. H. J. Daysh (ed.), *A Survey of Whitby* (Eton: 1958), pp. 66–70, 148–54; N. Pevsner, *Yorkshire, North Riding* (London: 1960), pp. 396–99; neg. no. BOQ 82, 29 July 1973.

66 Thorpeness, Suffolk

Thorpeness is a planned seaside holiday village for the middle classes. In 1908, Mr Glencairn Stuart Ogilvie (1858–1932), fifth son of Alexander Ogilvie (1812–80), unexpectedly inherited a large estate of some 2500 hectares on the Suffolk coast, built up by his father who had made a fortune from railway and dockland construction. The estate included an area of heathland and beach just north of Aldeburgh. In 1910 he set up a company called Seaside Bungalows Limited, later Thorpeness Limited, and began to lay out a village behind the beach to provide rented houses for self-catering holidays. Visitors arriving by train at a special halt built by the Great Eastern Railway on the branch line from Saxmundham were met by a brake and luggage car.

At first the future looked promising. A shallow lake, one metre deep and 27 hectares in extent, with numerous wooded islands within it, was created in a park-like setting behind the beach. In keeping with the middle-class ethos the islands, which could be rented by the day, were given names of characters in books by J. M. Barrie who was a friend of Ogilvie. These islands were provided with structures such as 'Wendy's House' and 'The Pirates' Lair'. The lake lies on the left of the picture which is taken from the east. On the north, right, of the lake two parallel roads were laid out but only one was developed. There, large detached houses, timber-framed and weatherboarded, were erected between 1911 and 1914. On the eastern, nearside, of the lake is a boathouse (1911) with its tower, originally serving afternoon teas. Smaller houses were built (1911–14) along the curving road to the north-east as well as a shop (1913) and a guest-house (1914). A country club just above the beach, centre, with tennis courts, was begun in 1914 and later enlarged.

After the First World War, although the resort became increasingly popular, the expected numbers of visitors failed to materialise and the company changed its policy. Much smaller timber-framed and cladded bungalows appeared (1919–20) along the northern of the two lakeside roads and were advertised as 'compact'. This allowed them to be cleaned by a maid who would then 'still be out on the beach with the mistress and children by 11.00 a.m.'. Another change was the selling of private plots. Later came very different buildings, mainly of brick. These included almshouses (1926) for retired estate workers and Westgate, a street of detached houses of Germanic appearance terminating in a towered gatehouse (1928–9), visible lower centre, above the beach. A clubhouse in a seventeenth-century French style, for the earlier golf course, top right, came in 1930. A vaguely Romanesque church and a tall curved sub art

deco terrace, bottom centre, overlooking the sea, completed the resort in 1936–9. After 1945 the popularity of Thorpeness declined and its great days were over.

M. Coleman, *Concerning Thorpeness* (Suffolk Preservation Society: 1984); N. Pevsner, *Suffolk* (London: 1974), pp. 464–5; A. Ogilvie de Mille, *One Man's Dream* (Dereham: 1996); neg. no. BSL 77, 20 July 1975.

67 Cathays Park, Cardiff

Cathays Park has been called 'the finest civic centre' (Newman) and the 'most rigidly geometrical park' (Elliott) in Britain. It is also a monument to both Welsh pride and Edwardian certainty. The site, here viewed from the north-west, a rectangular area of 24 hectares, was bought by the City in 1898. The grounds, known as The Alexandra Gardens, were laid out between 1904 and 1909 by William Harpur (fl. 1890s–1900s), the Borough Engineer, as a flat, unornamented, green open space bounded by broad tree-lined roads. Over the next sixty years a series of monumental buildings were erected around them. None of these buildings dominates the area and collectively they form a visual unity.

The first building was the City Hall (1901–5), a grand Edwardian baroque edifice, at the far end. To its right, in the corner, are the Law Courts (1901–4), in the same style. To the left, in the other corner, is the National Museum of Wales in a more restrained baroque style, begun in 1913 but not completed until 1927. Below the museum and only partly visible is the massive, classical, University of Wales, Cardiff, building (1905–9). In the foreground, facing south-east, is the Welsh Office (1934–8), still classical. To its right, in the bottom right corner, is the Temple of Peace and Health (1937–8), originally intended as a war memorial, and above it is the Bate Building (1913–16), once the Technical Institute. Next is the Mid-Glamorgan County Hall (1912) with the University Registry (1903–4) adjacent to it. All these are in a classical style. Finally, between the Registry and the Law Courts, is the Central Police Station (1966–8), the concrete elevations of which contrast sharply with those of its neighbours. In the roundel in the centre of the gardens is the Welsh National War Memorial (1924–8), by Sir Ninian Comper (1864–1960).

B. Elliott, *Victorian Gardens* (London: 1986), p. 238; J. Newman, *Glamorgan* (London: 1995), pp. 220–35; neg. no. BPJ 65, 25 January 1974.

68 Lichfield, Staffordshire

Here, in a narrow valley between the town, right, and the ancient cathedral, and viewed from the south-west, is a designed landscape with a long and complex history. The cathedral dates to before the seventh century and the town is an episcopal foundation of 1149. There may have been a natural pool on the site of the present one but it was dammed certainly before 1183 to provide power for a mill and was used as a fishpond by 1298. Another pool lay upstream to the west on the site of the formal gardens in the foreground. This was separated from the first pool by a causeway and bridge, the modern replacement of which can be seen between the trees. This bridge was described as 'new' in 1315. It is possible that the causeway, together with the pool itself, was intended to enhance the overall landscape setting of the cathedral which was virtually rebuilt between 1220 and 1330. In time the upper pool silted up and by 1700 all the land west of the causeway was marsh.

In the eighteenth century Lichfield became a minor provincial centre. It also, for its time, had an excellent water supply provided by the Conduit Lands Trust, who took over a medieval monastic system of water provision. Luck, and good management, produced a considerable surplus income for the Trust which was used for general benevolent purposes. As part of the improvements to the town, and encouraged by the local literary figure Anna Seward who wanted to replicate the Hyde Park Serpentine, the Trust improved the Pool and created a sinuous public walk along its southern edge.

In 1856 the new South Staffordshire Water Company acquired the Pool as a reservoir and it was dredged and revetted. In 1859 the Trust built a museum in an Italianate style on the western edge of the marshland. This is just visible in the top left-hand corner of the formal garden. The Water Company dumped the material dredged from the Pool on to this marsh, so creating land for a Museum Garden. This was laid out in 1860–1 in a formal style with a fountain in a central roundel and was again paid for by the Trust.

The next development was in 1920 when a Garden of Remembrance was created in the trapezoidal area of land to the left of the nearer end of the Pool. In the 1960s the Pool passed into the hands of the Local Authority who widened the southern edge to produce a more open walk. Thus was created, over 600 years, a public amenity of great value. This is the situation captured in the photograph. Since then, the inevitable car park south of the Pool and an 'inner ring' road along the southern edge of the garden have done little to enhance the landscape.

P. Laithwaite, *The History of the Lichfield Conduit Land Trust* (Lichfield: 1947); VCH, *Staffordshire* XIV (Oxford: 1990), passim; neg. no. SB 91, 6 April 1956.

69 Cumbernauld, Dunbartonshire

Cumbernauld lies 20 kilometres north-east of Glasgow. It is one of the first post-war Scottish New Towns which, following the 1946 New Towns Act, offered the hope of better living conditions for the inhabitants of many of Scotland's inner cities and, in particular, for those in the Glasgow slums. It was begun in 1956 and comprised a series of self-contained districts, linked by a ring road and served by a central spine route. It was given a mixture of housing types including the traditional Scottish flats or apartments, tower blocks and detached and terraced houses. It now has a population of over 50,000.

Here the view is from the north-east with the central road on the right and the south-western half of the inner ring road running roughly parallel to it on its left. The town looks more spectacular from the air than does Milton Keynes [57] but it is not so successful as a social entity. It has all the architectural brutality of the 1950s and 1960s, and the influence of the road engineer is obvious. Although many of the detached houses and terraces do have their own gardens, the flats and tower blocks stand in carefully designed open spaces, sometimes grassed, often paved. While there can be no doubt that Cumbernauld is a vast improvement on the Glasgow slums it replaced and is better than the earlier East Kilbride, to the south-east of Glasgow, it cannot be considered a success in the history of designed landscapes.

Neg. no. BGB 43, 9 July 1971.

70 Christ's Hospital, Sussex

Most schools, indeed most institutions, begin by endeavouring to have a pleasant landscape around them. But, almost inevitably, the conflict between the demands for physical recreation, the need for continual expansion to cope with changing educational requirements, and the preservation of the designed landscape, usually results in the last being the loser.

Christ's Hospital was founded by Edward VI in 1553 and established in the City of London on the site of the then newly dissolved House of Greyfriars in Newgate. As is the case with similar schools, it was relocated in the late nineteenth century on a greenfield site just south-west of Horsham and planned as a form of institutional country house. The new school was begun in 1893 to designs by Sir Aston Webb (1849–1930) and was completed in 1902. The view here is from the north-west and shows the intended park-like environment with copses, belts and isolated trees. The administrative and teaching part of the school, including the hall and chapel, lies in the middle of the photograph, set around a large court or quadrangle with a central fountain. To the north, left, of the quadrangle are the boarding houses, eight H-shaped blocks on a slight curve. These houses have small gardens, at least two of which were created by the wife of Sir Frank Fletcher, headmaster 1911–35, with the assistance of their friend, Gertrude Jekyll (1843–1932). The surrounding parkland is still reasonably intact but, inevitably, later buildings have encroached upon it and it includes numerous playing fields, pitches and a running track.

N. Pevsner, *Sussex* (London: 1965), pp. 187–8; J. Brown, *Gardens of a Golden Afternoon* (London: 1982), p. 197; neg. no. FD 34, 11 June 1950.

71 University of Exeter, Devon

The University College was founded in 1922, based on the Albert Memorial College, established in the 1860s, and housed in various buildings in Exeter. The first principal of the new college, Professor (later Sir) Hector Hetherington (1888–1965), had the foresight to encourage the College Council to purchase the Stretham Hall estate on the northern outskirts of the city, following the gift by H. W. Reed of the Hall itself together with its garden.

Stretham Hall, now Reed Hall, was built in 1867 in a heavy Italianate style. It and its surrounding estate were bought by Richard Thornton West (?–1878), a retired East India merchant. West employed the Exeter branch of the firm of Robert Veitch and Son, run by Peter Veitch (1850–1929), to plant an Italian terrace garden and to landscape the adjacent parkland. The result was a fine late-Victorian estate with many exotic specimen trees.

The initial development of the University College was on the Stretham estate but gradually spread beyond. New roads and buildings appeared, the latter including teaching and administrative blocks as well as residences. Despite this, the College put much effort into preserving and improving the appearance of the landscape. A number of other Victorian villas were acquired and the often densely planted gardens were incorporated into the college grounds. Further expansion took place after the Second World War and in 1956 the College was granted the status of a full university. As with most universities in the 1960s and 1970s, Exeter expanded considerably, more than trebling its student numbers. This growth, and the subsequent provision of new buildings, inevitably put pressure on the older designed landscape, not always to its benefit. However, the University has struggled to keep as much of the old parkland as possible and further landscaping has been carried out by Dame Sylvia Crowe (1901–97).

In essence, the campus has been developed as an extended arboretum, based on a small one begun by Veitch and Son and on some of the smaller Victorian gardens. The former includes an exceptional pinetum, the latter fine specimens of Lucombe oaks. The relatively mild climate has also allowed various sub-tropical trees to flourish. In the photograph, taken from the south-west, the wooded area in the left centre is part of the original Stretham Hall park, with the Hall centre left, while the broad curving drive is the Prince of Wales Road (1931). Beyond, some of the boundaries of the former fields are still visible between the later playing fields and buildings.

The Grounds and Gardens of Exeter University (Exeter: 1969); B. W. Clapp, *The University of Exeter* (Exeter: 1982); A. G. Hellyer, *Shell Guide to Gardens* (London: 1977), pp. 132–4; neg. no. CMN 26, 5 July 1980.

72 Crichton Hospital, Dumfriesshire

Crichton Hospital lies just south-east of Dumfries and is here seen from the north-west. Dr James Crichton (?–1823), who made a fortune with the East India Company, left £100,000 for charitable purposes in Dumfriesshire. In 1833 his trustees decided to spend the money on an asylum. They purchased 16 hectares of farmland and in 1835 work began to designs by William Burns (1789–1870). The hospital was opened in 1839 and has continued to expand ever since. Additional buildings were constructed in 1848–9, 1867–71, 1890–1912 and 1922–38, mostly in separate blocks around the spacious grounds.

The buildings, as well as the grounds, illustrate to perfection the architectural response to the changing attitudes, to and treatment of, the mentally ill over 150 years. The buildings include the Elizabethan-style original institutional ones by Burns, enlarged by W. L. Moffatt in 1867–70, visible in the foreground, the huge Gothic church (1890–7) by S. Mitchell in the centre, the Royal Farm and patient accommodation (1890–2) in the far right background, and a series of large detached houses (1898–1912) for different categories of patient, arranged in an arc beyond the church. After the First World War came a hospice (1912–27), the long range just right and above the original building, a nurses' home (1922–3) near the farm, more individual patient accommodation (1923–4), all in a neo-Georgian style, and the large therapeutic and recreational centre (1934–8), upper left, in an art deco style. There is also a bowling green and a pavilion (1894), an electrical power station (1894–5) and a sports pavilion (1923).

The grounds were well wooded from the beginning, with many hedgerow trees from the earlier fields being retained, as can be seen from their straight alignments. Further planting continued throughout the nineteenth century, as did an increasingly complex arrangement of drives. The building programme of the 1920s was accompanied by further tree and shrub planting, as well as by the creation of a number of separate gardens around the patient accommodation houses and, clear in the photograph, alongside the new hospice. The designs for these were by Sir George Watt (1851–1930). Watt was a keen botanist and former Indian civil servant who had retired to Lockerbie in 1906. The new gardens were provided with cuttings and seeds by local people, from the Royal Botanic Garden, Edinburgh, and, presumably because of Watt, from India.

J. Gifford, *Dumfries and Galloway* (London: 1996), pp. 253–63; neg. no. CAM 83, 27 July 1976.

73 Botanic Gardens, Cambridge

Several attempts were made to make a botanic garden in Cambridge to match that in Oxford (1621) but none was successful until 1762 when one was established near the city centre, by no means an ideal position. In 1831 the University acquired the present site, covering 16 hectares, then well outside the town. The initial plan was to develop all the land but, in the end, only the western half, nearest the camera in this view from the west, was laid out between 1846 and 1860. It was designed by J. S. Henslow (1796–1861), Professor of Botany at Cambridge (1825–60), and contains the reference collection of all the notable trees native to Europe and North America, including some varieties rare in Britain. To the north, left, of the main axial walk is a small lake, a stream, the Water Garden and the Rock Garden. The latter contains a section of alpine plants arranged by continent. To the right of the main walk the elaborately patterned arrangement of the Systematic Beds is visible. This is an important teaching collection of hardy flowering plants and the beds indicate botanical relationships as well as being decorative. Above and half-left of the Rock Garden is a range of glasshouses which hold collections of temperate, alpine, tropical and semi-arid plants.

The eastern half of the garden was not developed until after 1951. Then a combination of a large bequest by Reginald Cory [58], already a great benefactor, the expansion of research by the School of Botany and the increasing demands of conservation, education and amenity, led to a gradual expansion, not complete when this photograph was taken. The new features included a research area, visible as narrow plots in the centre background, a small ecological area of native British plants in their natural habitats, a scented garden for the blind, a collection of hybrids and their parents and a picnic area for visitors. Since the photograph was taken a Winter Garden, a Chronological Garden planted with shrubs and herbaceous plants in order of their introduction into Britain, a rose garden and a pinetum have been added. Buildings on the site include teaching facilities, laboratories, a library and an exhibition and education centre. Today the gardens successfully combine the roles of teaching, research and amenity.

Cambridge University Botanic Garden Visitors' Guide (Cambridge: 1991); H. Gilbert-Carter, *Guide to the University Gardens* (Cambridge: 1947); F. C. Preston, 'University Botanic Garden, Cambridge', *Journal of the Royal Horticultural Society* 65 (1940), pp. 171–3; VCH, *Cambridgeshire* III (London: 1959), pp. 324–5; neg. no. AIU 10, 25 May 1964.

Chapter 10

The Future of Past Gardens

Those who attempt to both write the history of parks and gardens and to influence their preservation soon find themselves in a philosophical dilemma. All historians study the past through written records but many also use structures or objects as primary evidence. Art historians use paintings, carvings and sculpture. Architectural historians use buildings. Military historians look at battlefields, weapons and defensive works. Medieval economic historians sometimes study ancient field systems or deserted villages. Garden historians are no different. They too use written records and illustrations as well as the parks and gardens that survive in various forms. But, in using gardens as a historical source, the garden historians are analysing evidence which is no longer in its original form, is changing all the time and which will continue to change, whether carefully managed or abandoned. Further, it can never be recreated 'as it was', even if that were desired. Although this also pertains to some extent in, for example, architectural and military history, it is more serious for garden historians and, together with the inexorable progress of Nature, is unusual. To have source material continually and rapidly changing is an interesting and somewhat intimidating historical problem.

The situation is made more difficult by the fact that garden historians, well understanding the historical, educational, social and aesthetic value of this basic evidence, are often involved in attempts to protect, conserve or restore parks and gardens. This is, after all, what lies behind much of the work of The Garden History Society, the County Garden Trusts, the Historic Parks and Gardens section of English Heritage and indeed many local authorities. Yet, complete protection or preservation is destined to fail. Any 'preserved' park or garden will in the end be only a pale reflection of its earlier form. Even the best historic garden is actually preserving something that was never there. And, as was noted in Chapter 8, true recreation or restoration is probably impossible.

There are also the same pressures on parks and gardens as on all historic landscapes in the overcrowded Britain of today. The various political ideologies, the insatiable demands for more roads, more houses and cheap food and the decline of that part of society which could afford to create and maintain extensive private parks and gardens means that such landscapes will always be under threat. Regardless of good will, dedication or even legislation, there will inevitably be constant attrition of the diminishing stock of historic parks and gardens. New roads and motorways have been cut through parkland and sports facilities added, as at Tredegar [74]. Other parks have been divided between various owners and ploughed up, as at Appuldurcombe [75], or built over as at Great Barton [76]. Elsewhere, rare and wonderful remains of gardens, as well as the evidence for other aspects of their past, have been destroyed, as at Eastbury [78] and Stainfield [79].

Despite the destruction of and, more especially, the inevitable changes to parks and gardens, much has survived. Many landowners have held on to their estates and protected their important parks and gardens in the face of considerable financial difficulties, even if they have had to adapt them to twentieth-century demands. The owners of Woburn [77] and Port Lympne [55] have introduced attractions which have ensured the preservation of the major part of the parks and gardens there. Elsewhere, hundreds of parks and gardens have been opened to the public on a less ambitious basis. Levens [27] and Somerleyton [45] are examples.

Local authorities too have responsibility for public parks such as Calverley Park [63], Cathays Park, Cardiff [67] and Princes Street Gardens, Edinburgh [64], while some have saved major former private gardens, as is the case with Hestercombe [62] and Somerset County Council. Elsewhere special trusts have been set up to protect and encourage the visiting of important designed landscapes. These landscapes include Leeds Castle [11] and Painshill Park, Surrey. The National Trust and the National Trust for Scotland hold over 200 gardens of varying size and importance, including Studley Royal [31], Wimpole [49], Penrhyn [50] and Pitmedden [61]. English Heritage, Historic Scotland and Cadw also have a number of fine gardens such as Osborne [47], Witley [46], Wrest Park [29], Raglan [20] and Edzell Castle, Angus.

Whether Britain will acquire equally fine gardens and designed landscapes in the next millennium is uncertain. The diversity of gardens, whether suburban plots, Chelsea Flower

Show set pieces or the landscapes of internationally recognised designers, is now so great that garden historians of the future may have some difficulty in tracing a coherent thread of development. It is perhaps landscape historians who will sit most comfortably in such a world. By looking at and analysing gardens and parks in the same way as they consider fields, villages, towns, roads, houses and castles, landscape historians will still be studying the same wonderful kaleidoscope that is the British landscape [80]. No historian could ask for more.

74 Tredegar Park, Monmouthshire

Tredegar Park lies immediately south-west of Newport and what remains is here seen from the south. The Morgan family lived at Tredegar from the fifteenth century and grew rich from the coal that underlay much of their land. As the leading Whig family in the county they also represented Monmouthshire in parliament from the 1660s until the early nineteenth century.

An old sixteenth-century house and garden were swept away by Sir William Morgan (c. 1640–80) who, between 1664 and 1672, built the fine brick mansion there that still survives. He also turned the gardens into fashionable Dutch ones. Morgan's fourth son, John (1671–1720), later created a baroque-style parkland with long straight avenues. The remnant of one of these is visible in the photograph, extending right to left. In 1790 John Morgan (?–1792) employed the Yorkshire landscape designer Adam Mickle (c. 1730–1809) to make the park less formal. Most of the avenues were broken up and the parkland in front of the house opened into a wide area of grass with extensive tree-planting beyond. A substantial lake was also created.

Morgan died without male heirs and Tredegar passed by marriage to Charles Gould, later Sir Charles Morgan (1726–1806), MP and lawyer. The family's wealth increased in the nineteenth century as the South Wales coalfield expanded. Charles Morgan's grandson, another Charles (1792–1883), was created Baron Tredegar in 1859 and his second son, Godfrey (1831–1913), a distinguished soldier, became a viscount in 1905. During these years the park was carefully maintained and pleasure gardens and a fine collection of conifers were added. However, death duties and especially the extravagant lifestyle of Evan Morgan (1893–1949), the second viscount, led to the sale of the estate in 1951. The house was eventually saved by Newport Borough Council who have created a late-seventeenth-century garden there. The park was not so fortunate. It was first used in part for playing fields and then the M4 motorway was driven through it with Junction 28 carefully placed across the last remaining early-eighteenth-century avenue.

D. Jacques, *Georgian Gardens* (London: 1983), pp. 116–17; E. Whittle, *The Historic Gardens of Wales* (London: 1992), pp. 4, 25–6, 32, 36, 48, 64; neg. no. CPI 28, 16 October 1981.

75. *Appuldurcombe, Isle of Wight*

Here, viewed from the south, is a great landscape park which has been torn apart. Yet, with its steep western side it must have been more imposing than most parks in lowland Britain.

Appuldurcombe belonged to the Worsleys, a leading Isle of Wight family who had lived there since the sixteenth century. In 1701–3 a new house and a garden were created there for Sir Robert Worsley (1686–1742). The house, now ruinous, stands in the trees in the middle distance. A 'new' park is documented in 1707 and certainly one with an early-eighteenth-century arrangement of long formal avenues existed by 1769. On Sir Robert's death the estate passed to a cousin and on to his son, Sir Richard Worsley (1751–1805). In 1779 Capability Brown visited Appuldurcombe and later produced a new plan for the park. What Brown did is not clear. He seems to have removed the old gardens from around the house and to have enlarged the park, but the expected lake did not materialise. With only one small stream on the eastern edge of the park, it may be that a lake was impossible to construct. A number of eye-catchers, including a triumphal arch by Wyatt, and an obelisk were erected.

Sir Richard did not enjoy his creation long. In 1782, after a scandal, he divorced his wife and left the country. He returned in 1797 but lived alone with his 'housekeeper' in a seaside cottage until his death. Appuldurcombe then passed by marriage to Charles Pelham (1781–1846), first Baron Yarborough of Brocklesby, Lincolnshire. Pelham used the estate as a base for his social activities as founder and first Commodore of the Royal Yacht Squadron. He seems to have created a formal garden and a wooded pleasure ground around the house which survive in part. After Pelham's death the estate was sold and the park divided. The house lingered on as a hotel, a school and a monastery, but was abandoned in 1907. It was later taken over as a ruin and is now in the care of English Heritage. The park was not so fortunate. Little of it remains except for some of the boundary plantations on the north and west and some isolated trees.

V. Basford, *Historic Parks and Gardens of the Isle of Wight* (Newport: 1989), pp. 32–4; R. Worsley, *History of the Isle of Wight* (London: 1781), p. 218; neg. no. COA 26, 15 April 1981.

76 Great Barton Park, Suffolk

Here, 3.5 kilometres north-east of Bury St Edmunds and viewed from the south-east, is an eighteenth-century landscaped park redesigned in the twentieth century for a different purpose. Its early history is not clear, but in the late seventeenth century the estate passed by marriage from the Folkes family to Sir Thomas Hanmer (1677–1746) and then to his sister's husband Sir William Bunbury (c. 1710–64). Sir William was a member of a distinguished Chester family but as vicar of Mildenhall, Suffolk, had little use for Great Barton. His son, Sir Thomas Bunbury (1740–1821), however, did. When he married in 1762 he settled at Great Barton and became MP for Suffolk. It was probably he who built a new mansion house there and laid out the small park of under fifty hectares around it. It had extensive plantations on its northern and western sides, pleasure grounds around the hall, a scatter of copses and individual trees in the centre and a long sinuous drive across it from the south-west. No designer is known and it is likely that, as with many similar undistinguished parks **(34)**, it was laid out by the owner himself. Sir Thomas had no direct interest in the landscape except as a setting for his house and for his social activities. He was passionately involved in racing, co-founder of, and owner of the first winner of, The Derby, and also gave his name to one of the courses at Newmarket.

The Bunburys lived at Barton until 1913 when the house was destroyed by fire. The north-western part of the park was subsequently returned to agriculture, but the southern third survived into the 1960s when the whole of the original park was developed as a select middle-class housing estate with a pseudo-village green. The hall stood in the trees at the centre right of the photograph. The north-west side of the 'green' follows the line of the old drive which continued north-eastwards along what is now a modern street. The copses on the 'green' and most of the mature individual trees are also part of the original park, as are the more extensive plantations on the north-north-west edges of the estate.

Journal of the British Archaeological Association 20 (1914), p. 77; Ordnance Survey 1st edn one-inch map, sheet 50 (1837); neg. no. BJI 91, 7 July 1972.

77 *Woburn Park, Bedfordshire*

This photograph shows another result of the pressures of the twentieth century on historic parkland. Woburn Park lies halfway between Luton and Milton Keynes and is one of the largest in Britain, covering 1200 hectares. In this view only the north-western half is visible, with Woburn Abbey, its pleasure grounds and the rest of the park lying behind the camera.

Although the estate was acquired by the Russell family in the sixteenth century it was not until the early seventeenth century that the earls of Bedford lived at Woburn. Thereafter, almost every successive Russell altered or improved the house and park. Bridgeman, Chambers, Holland and especially Repton have all left their mark there. From the early seventeenth century to the mid-eighteenth century the park was less than half its present size and lay entirely south of, below, the road running across the lower right of the photograph. The exact dates of its extension are unknown. The area immediately north of the road was emparked between 1760 and the 1780s but the extreme northern part was still not enclosed when the first edition Ordnance Survey one-inch map of 1834 was published and no landscape designer has been credited with the work.

On the death of the twelfth duke in 1953 the enormous, at that time, sum of five million pounds demanded in death duties required drastic action. To keep the estate intact both house and park had to be exploited commercially. Much of the northern part of the park was turned into a safari park. This comprises large fenced areas containing wild animals, through which visitors drive along the sinuous roads visible in the photograph. Later an extensive area with amusements of various types was added, seen in the centre-left middle distance. The result may not be visually attractive, least of all from the air, but it has helped to ensure the survival of a magnificent designed landscape and its associated buildings.

P. Bigmore, *The Bedfordshire and Huntingdonshire Landscape* (London: 1979), pp. 147–8; G. Carter, P. Goode and K. Laurie, *Humphrey Repton* (London: 1982), p. 60; N. Pevsner, *Bedfordshire, Huntingdonshire and Peterborough* (London: 1966), pp. 166–71; neg. no. BMF 53, 25 April 1973.

78 Eastbury, Dorset

This photograph shows the last traces of part of one of the greatest early-eighteenth-century gardens in Britain. The site, seven kilometres north-east of Blandford, was a new one when, in 1718, George Doddington (?–1720) commissioned Sir John Vanburgh (1664–1726) to build a house here. Doddington, a member of an old Somerset family and holder of a number of lucrative government offices including Treasurer of the Navy and a Commissioner of the Admiralty, had no direct heir. He therefore arranged that George Bubb (1691–1762), the son of his sister and her husband, a Weymouth apothecary, should inherit his estate on condition that Bubb took his uncle's name.

The house was unfinished on George Doddington's death, but work on it began again in 1722, simultaneously with the creation of a park and garden, designed by Charles Bridgeman (?–1738) and said to be 'the most authentic notion of his early style'. George Bubb Doddington became a flamboyant and eccentric figure who spent much of his later life in political intrigue. Considerable sums of money were used to ensure the election and control of the MPs for his home towns of Melcombe and Weymouth. Long before it was completed in 1738 Eastbury had become well known for its literary and artistic visitors, Fielding, Thornhill and Voltaire among them. Doddington died unmarried in 1762, a year after being created Lord Melcombe. The house was largely demolished between 1775 and 1782 and the gardens abandoned.

The formal gardens lay behind the house. The main compartment had an axial vista from the house across a lawn and then along a canal. This canal was edged by elaborate parterres with further complex parterres outside them which, on the north, had two large octagonal mounds. The vista continued through a wilderness of rectangular compartments with central *rondes* and radiating *allées* and then into a shallow valley containing a circular basin. This basin was backed on the rising ground by a rectangular amphitheatre surmounted by a Corinthian Temple designed by Vanbrugh.

The whole garden survived, albeit as an archaeological site, until the 1960s. Then its entire eastern end, as well as much of the park, was destroyed by ploughing. In this view, from the south-west, the last traces of the basin and amphitheatre are visible as cropmarks.

J. Collinson, *History and Antiquities of Somerset* III (Bath: 1791), pp. 518–19; C. Hussey, *English Gardens and Landscapes* (London: 1967), p. 38; RCHME, *Dorset* IV (London: 1972), pp. 90–3; L. Whistler, 'Eastbury Park', *Country Life* 104 (1948), pp. 1386–9; neg. no. ANC 22, 12 March 1966.

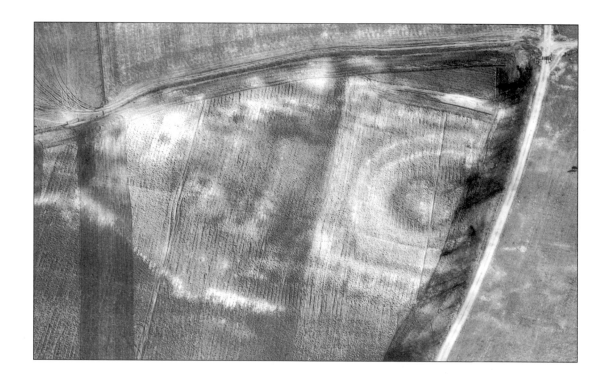

79 Stainfield, Lincolnshire

Here is where garden and landscape history meet. The view looking south-east is of a single pasture field 15 kilometres east of Lincoln. The story of the landscape begins with the late Saxon village of Stainfield which lay close to a stream in the foreground. In the early twelfth century this village was cleared away and a priory of Benedictine nuns established on its site. The priory was enclosed by a large bank and ditch on three sides and part of this boundary is visible running south-west, right, of the modern house. The church and conventual buildings were near this house, while fishponds, a mill and various outbuildings, some of which survive as earthworks, lay on the site of the earlier village, in the foreground. The dispossessed villagers were resettled in a new planned village, laid out on the rising ground to the south-east of the priory. Part of the remains of this village can be seen in the top right of the picture as ditched closes and former house sites with a fragmentary hollow-way, once the main street, on their far side. The village remains extended further south-east on the other side of the street but they were destroyed by ploughing before the photograph was taken.

The priory was dissolved in 1536 and in its ruins Sir Philip Tyrwhitt (?–1588) built a country house. His grandson, another Sir Philip (c. 1560–1624), enlarged the house and created a park, probably soon after 1611 when he was made a baronet. He also laid out a formal walled garden south of his house, seen here in the top left of the photograph cutting through part of the remains of the village. The village itself was cleared away yet again as part of the emparking. The garden remains comprise two parts. First, an L-shaped raised terrace overlooking a rectangular compartment crossed by an axial path and with two further compartments to the east. Second, above it, a rectangular ditched area overlying medieval plough ridges and part of a former village lane, which was presumably planted as a 'wilderness'.

The house and garden remained unaltered throughout the seventeenth century but in 1707–10 Sir John Tyrwhitt (c. 1660–1741) refurbished the hall and, while leaving the garden untouched, laid out a formal landscape to the west. Some features of this landscape are visible as broad low banks, set within the old monastic precinct, splaying outwards to the west and with a central avenue. This afforded vistas of Lincoln Cathedral on the horizon and of the ruins of Barlings Abbey two kilometres away. The last of the Tyrwhitts died in 1760, the house was abandoned and the site became a working farm. The earthwork evidence for 1,200 years of history was destroyed by ploughing in 1980.

P. L. Everson, C. C. Taylor and C. J. Dunn, *Change and Continuity* (London: 1991), pp. 175–8; neg. no. CFJ 24, 10 February 1978.

80 Croxton Park, Cambridgeshire

Croxton Park, 20 kilometres west of Cambridge and here viewed from the west, is an example of landscape rather than of garden history. It is an undistinguished piece of parkland which nevertheless displays its past. A small park, occupying little more than an area in the upper centre and left of the photograph, was created in 1761 by Edward Leeds (1728–1823). Leeds was a member of a minor landed family, most of whom were lawyers, which had owned Croxton since the late sixteenth century. Sometime in the 1570s the medieval manor house was rebuilt. At the same time gardens, which included a curious and possibly unique circular moated layout with a raised walk and outer western canal, were created. The remains of this garden can be seen in the photograph, upper centre. These were abandoned and incorporated in the new park in 1761. The house, seen here above the garden remains, was rebuilt at the same time.

In 1825 the last of the Leeds sold Croxton to Samuel Newton (?–1848), the grandson of an upwardly mobile Liverpool merchant. The house was suitably grand but the park was not. Newton therefore cleared away what remained of Croxton village, which lay beyond the church and the manor house, out of the picture, and laid out a much larger park. Not all of this park was newly planted. As so often happens, existing hedgerow trees were incorporated into the new park and the old hedge banks were left. Both of these features can be seen in the photograph, especially bottom left. The creation of this nineteenth-century park not only preserved the recently abandoned streets, lanes, paddocks and house sites of the village, but also protected the remains of part of the adjacent village of Westbury which, by then, had declined to less than half of its fourteenth-century size. In the foreground, just below the drive to the Hall, once a medieval street, are the abandoned peasant houses of Westbury. Elsewhere are the remains of the ridge-and-furrow of the medieval fields of both Croxton and Westbury. Although as a park it is not of the highest quality, Croxton contains one of the best-preserved medieval landscapes in East Anglia.

A. E. Brown and C. C. Taylor, 'Cambridgeshire earthwork surveys VI', *Proceedings of the Cambridge Antiquarian Society* 82 (1993), pp. 101–8; RCHME, *West Cambridgeshire* (London: 1968), pp. 63–71; neg. no. CHX 5, 31 January 1979.

Notes to Chapters

Chapter 1

1. D. R. Wilson, *Air Photographic Interpretation for Archaeologists* (London: 1982).
2. P. Everson, 'Bodiam Castle', *Château Gaillard* 17 (1996), pp. 79–84.
3. J. Heward and R. Taylor, *The Country Houses of Northamptonshire* (London: 1996), pp. 31, 66, 173, 268.
4. L. Knyff and J. Kip, *Britannia Illustrata* (London: 1707, reprinted 1984); G. Jackson-Stops, *An English Arcadia* (London: 1991), pp. 11–13.

Chapter 2

1. W. Torbrügge, 'Vor- und frühgeschichtliche Flussfunde'. *Ber. Rom.-Ger. Komm. 1970–1* (1972), pp. 1–146; J. May, *Prehistoric Lincolnshire* (Lincoln: 1976), pp. 129–32, 165–7.
2. T. Gregory, 'Excavations in Thetford 1980–2', *East Anglian Archaeology* 53 (1991), pp. 196–201.
3. B. Cunliffe, *Excavations at Fishbourne* I, Society of Antiquaries of London Research Report (London: 1971).
4. R. J. Zeepvat and R. J. Williams, *Bancroft*, Buckinghamshire Archaeological Society Monograph 7 (1) (1994), pp. 129–215; H. G. Gracie and E. G. Price, 'Frocester Court Villa: Second Report', *Transactions of the Gloucestershire and Bristol Archaeological Society* 97 (1979), pp. 9–64.
5. J. Wacher, *The Towns of Roman Britain* (London: 1974), pp. 91–3.
6. B. Radice, *Letters of Pliny the Younger* (Oxford: 1969), II, p. 17; C. Thacker, *The History of Gardens* (London: 1979), pp. 18–25.
7. R. Agache, 'New aerial research in Picardy and Artois', *Antiquity* 46 (1972), pp. 117–23.
8. W. J. and K. A. Rodwell, *Rivenhall*, CBA Research Report 4 (London: 1986), pp. 43–5.

Chapter 3

1. J. Harvey, *Medieval Gardens* (London: 1981), pp. 25–38.
2. Harvey 1981, pp. 52–93.
3. C. Thacker, *The Genius of Gardens* (London: 1994), p. 31.
4. P. Rose, 'The medieval garden at Tintagel', *Cornish Archaeology* 33 (1994), pp. 170–82.
5. A. E. Brown and C. C. Taylor, 'A relict garden at Linton', *Proceedings of the Cambridge*

Antiquarian Society 80 (1991), pp. 62–7; J. Bond, 'Beckley Park', *Oxford Local History* 3.1 (1988), pp. 1–13 and 3.2 (1989), pp. 47–61.

6. Harvey 1981, p. 85.
7. Harvey 1981, p. 83.
8. C. C. Taylor, 'Spaldwick', *Proceedings of the Cambridge Antiquarian Society* 78 (1989), pp. 71–5.
9. J. N. Hare, *Bishops Waltham Palace*, English Heritage Guide (London: 1987), p. 31.
10. L. Cantor, *The Medieval Parks of England* (Loughborough: 1983).
11. A. R. Myers (ed.), *English Historical Documents* IV (London: 1969), pp. 1157–8.
12. Royal Commission on the Historical Monuments of England, *Northamptonshire* I (London: 1975), pp. 50–1; *Northamptonshire* VI (London: 1984), p. 87; Dr R. Hoppitt, personal communication.
13. E. Roberts, 'Edward III's lodge at Odiham', *Medieval Archaeology* 39 (1995), pp. 91–100.
14. P. Stamper, *Historic Parks and Gardens of Shropshire* (Shrewsbury: 1996), p. 8.
15. P. L. Everson, C. C. Taylor and C. J. Dunn, *Change and Continuity* (London: 1991), p. 185; M. Aston, 'The earthworks at the Bishop's Palace, Alvechurch', *Transactions of the Worcestershire Archaeological Society* 3rd ser. 3 (1970-2), pp. 55–9; Hare 1987, p. 31; C. C. Taylor, 'From recording to recognition' in P. Pattison (ed.), *The Field Archaeology of Parks and Gardens* (London: 1998).
16. Everson 1996, pp. 79–84.
17. Harvey 1981, pp. 10, 80, 114.
18. Harvey 1981, pp. 10, 28; A. H. Van Buren, 'Reality and literary romance in the park at Hesdin' in E. B. MacDougall (ed.), *Dumbarton Oaks Colloquium on the History of the Landscape* 9 (1986), pp. 117–34.
19. Harvey 1981, p. 77.
20. RCHME 1975, pp. 30–1.
21. Everson et al. 1991, p. 185.
22. J. Munby, *Stokesay Castle*, English Heritage Guide (London: 1993)

Chapter 4

1. R. Strong, *The Renaissance Garden in England* (London: 1979); Thacker 1994, chapters 4–8.
2. Strong 1979, p. 49.
3. RCHME, *Northamptonshire* III (London: 1981), pp. 105–9; A. E. Brown and C. C. Taylor, 'Cambridgeshire earthwork surveys', *Proceedings of the Cambridge Antiquarian Society* 68 (1978), pp. 59–75.
4. A. E. Brown and C. C. Taylor, 'The gardens at Lyveden', *Archaeological Journal* 129 (1973), pp. 154–60; P. Everson, 'The gardens at Campden House', *Garden History* 17.2 (1989), pp. 109–21; RCHME 1981, pp. 105–9.

5. Everson et al. 1991, frontispiece, pp. 66–9.
6. Everson et al. 1991, pp. 184–5.
7. M. Aston (ed.), *Medieval Fish, Fisheries and Fishponds*, British Archaeological Report 182 (Oxford: 1988).
8. C. C. Taylor, 'Medieval moats in Cambridgeshire' in P. J. Fowler (ed.), *Archaeology and the Landscape* (London: 1972), pp. 237–48.
9. Brown and Taylor 1978; RCHME, *West Cambridgeshire* (London: 1968), p. 47.
10. RCHME 1981, plates 16, 17.
11. M. Airs, *The Tudor and Jacobean Country House* (Stroud: 1995), chapter 1.
12. H. Boyle, 'Elizabethan entertainment at Elvetham', *Studies in Philology* 68 (1971), pp. 146–66.
13. Thacker 1994, pp. 71–3, 137.

Chapter 5

1. Strong 1979, pp. 138–68.
2. J. D. Hunt, *Garden and Grove* (London: 1988), pp. 3–72; W. E. Mead, *The Grand Tour in the Eighteenth Century* (New York: 1914); Thacker 1979, pp. 95–112, 169–73.
3. Thacker 1979, pp. 139–62.
4. P. Pattison, 'Giant steps' in Pattison (ed.) 1998.
5. D. Jacques and A. J. Van de Horst, *The Gardens of William and Mary* (London: 1988).
6. G. Jackson-Stops, 'A formal garden reformed', *Country Life* 154 (1973), pp. 864–6; G. Jackson-Stops, *Westbury Court Gardens*, National Trust Guide (London: 1984).
7. S. Hurley (ed.), *The Restoration of the King's Privy Garden at Hampton Court* (London: 1995).
8. T. Williamson, *Polite Landscapes* (London: 1995), pp. 48–77.
9. Williamson 1995, pp. 52–7.
10. Williamson 1995, pp. 15–18; J. Brewer, *The Pleasures of the Imagination* (London: 1997).
11. A. Taigel and T. Williamson, 'Some early geometric gardens in Norfolk', *Journal of Garden History* 11 (1991).
12. W. G. Hiscock, *John Evelyn and His Circle* (London: 1995); J. Harvey, *Early Nurserymen* (Chichester: 1974), pp. 39–57.
13. Brown and Taylor 1973; C. C. Taylor, 'The Wothorpe landscape', *Proceedings of the Cambridge Antiquarian Society* 85 (1997), pp. 161–70; M. Girouard, *Life in the English Country House* (Yale: 1978), pp. 106–9.

Chapter 6

1. D. Jacques, *Georgian Gardens* (London: 1983); Williamson 1995.
2. J. D. Hunt, *William Kent* (London: 1987); M. Wilson, *William Kent* (London: 1984).

3. D. Stroud, *Capability Brown* (London: 1965); R. Turner, *Capability Brown and the Eighteenth-century English Landscape* (London: 1985).

4. G. and S. Jellicoe (eds), *Oxford Companion to Gardens* (Oxford: 1991), pp. 333, 470, 613; F. Cowell, 'Richard Woods', *Garden History* 14.2 (1986), pp. 85–119.

5. Harvey 1974, pp. 90–108.

6. D. Stroud, *Humphrey Repton* (London: 1962); G. Carter, P. Goode and K. Laurie, *Humphrey Repton* (Norwich: 1982).

7. J. Appleton, 'Some thoughts on the geography of the Picturesque', *Journal of Garden History* 6.3 (1986), pp. 270–91; G. Daniels and C. Watkins, 'Picturesque landscapes and estate management', *Rural History* 2.2 (1991), pp. 141–70.

8. Williamson 1995, pp. 160–5.

9. G. Sheernan, *Landscape Gardens in West Yorkshire* (Wakefield: 1990), p. 54.

Chapter 7

1. B. Elliott, *Victorian Gardens* (London: 1986).

2. Elliott 1986, figs 62, 64, 83, 91.

3. H. Repton, *The Landscape Gardens of Humphrey Repton . . . by J. C. Loudon* (London: 1840).

4. Elliott 1986, p. 62.

5. B. Elliott, 'Master of the geometric art', *Journal of the Royal Horticultural Society* 106.12 (1981), pp. 24–9.

6. D. Watkins, *Thomas Hope* (London: 1968).

7. J. Brown, *The English Garden in Our Time* (Woodbridge: 1986), pp. 12–38; W. Robinson, *The English Flower Garden* (London: 1883, reprinted 1996); M. Tooley and P. Arnander (eds), *Gertrude Jekyll* (Witton-le-Wear: 1995).

8. Elliott 1986, pp. 37–9.

9. M. Batey and D. Lambert, *The English Garden Tour* (London: 1990), pp. 286–90; O. Garrett, *Biddulph Grange Gardens*, National Trust Guide (London: 1992).

Chapter 8

1. D. Ottewill, *The Edwardian Garden* (London: 1989).

2. G. Jellicoe, *Guelph Lectures on Landscape Design* (Ontario: 1983).

3. P. Shepheard, *Modern Gardens* (London: 1953).

4. Brown 1986, pp. 111–34; C. Tunnard, *Gardens in the Modern Landscape* (London: 1938).

5. Brown 1986, pp. 85–110.

6. P. Oliver, I. Davis and I. Bentley, *Dunroamin* (London: 1981), pp. 67–9.

7. R. Gradidge, *Dream Houses* (London: 1980).

8.   A. A. Jackson, *Semi-detached London* (2nd edn, Didcot: 1991), pp. 124–5.

9.   Oliver et al. 1981, pp. 11–14.

10.   E. Clarke, *Hidcote* (London: 1989).

11.   D. H. Binney, *A Short History of Kiftsgate Court* (n.d.); K. Lemmon, *The Gardens of Britain* 5 (London: 1978), pp. 93–7; G. S. Thomas, *Gardens of the National Trust* (London: 1979), pp. 226–7.

12.   S. Nicholson, *Nymans* (Stroud: 1994); R. King, *Tresco* (Salem: 1985); Thomas 1979, pp. 245–9; T. Wright, *The Gardens of Britain* 4 (London: 1978), pp. 27–9.

13.   E. Banks, *Creating Period Gardens* (Oxford: 1991); S. Landsberg, *The Medieval Garden* (London: 1996), pp. 101–30; R. Strong, *Small Period Gardens* (London: 1992).

14.   J. Harvey, *Restoring Period Gardens* (Aylesbury: 1988); Hurley 1995.

Chapter 9

1.   D. Jacques, 'The chief ornament of Grays Inn', *Garden History* 17.1 (1989), pp. 41–67.

2.   C. C. Taylor, *Dorset* (London: 1970), pp. 199–200.

3.   N. Pevsner, *London* 1 (London: 1957), p. 587; *London* 2 (London: 1952), pp. 214–15.

4.   S. Lasdun, *The English Park* (London: 1991), pp. 129–31.

5.   H. Conway, *Public Parks* (Aylesbury: 1996); H. Jordan, 'Public parks 1895–1915', *Garden History* 22.1 (1994), pp. 85–113; Lasdun 1991, pp. 135–202.

6.   M. Aston and J. Bond, *The Landscape of Towns* (London: 1976), pp. 177–8; J. Simmonds, *The Railway in Town and Country* (Newton Abbot: 1986), pp. 235–68.

7.   Brown 1986, pp. 219–40.

8.   Mr R. Taylor, personal communication.

9.   E. F. Laidlaw, *The Story of the Royal National Hospital Ventnor* (privately published: 1980).

10.   S. E. Large, *King Edward VII Hospital 1901–80* (Midhurst: 1986).

Index